Eureka Math
Grade 5
Modules 5 & 6

Special thanks go to the Gordon A. Cain Center and to the Department of Mathematics at Louisiana State University for their support in the development of *Eureka Math*.

For a free *Eureka Math* Teacher
Resource Pack, Parent Tip
Sheets, and more please
visit www.Eureka.tools

Printed in the U.S.A.
This book may be purchased from the publisher at eureka-math.org
10 9
ISBN 978-1-63255-310-2

Name _____ Date _____

1. Use your centimeter cubes to build the figures pictured below on centimeter grid paper. Find the total volume of each figure you built, and explain how you counted the cubic units. Be sure to include units.

A.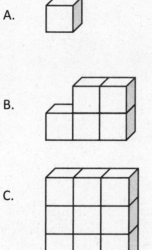

D.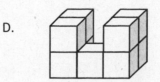

B.

E.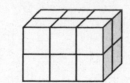

C.

F.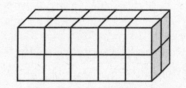

Figure	Volume	Explanation
A		
B		
C		
D		
E		
F		

Lesson 1:   Explore volume by building with and counting unit cubes.

1

2. Build 2 different structures with the following volumes using your unit cubes. Then, draw one of the figures on the dot paper. One example has been drawn for you.

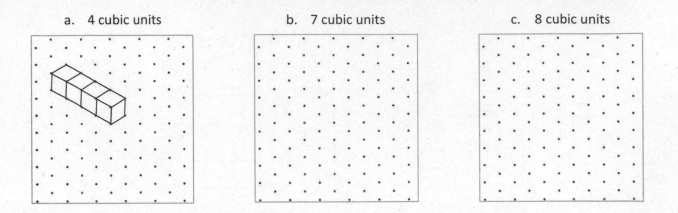

a.   4 cubic units   b.   7 cubic units   c.   8 cubic units

3. Joyce says that the figure below, made of 1 cm cubes, has a volume of 5 cubic centimeters.
    a.   Explain her mistake.

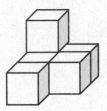

    b.   Imagine if Joyce adds to the second layer so the cubes completely cover the first layer in the figure above. What would be the volume of the new structure? Explain how you know.

Lesson 1:   Explore volume by building with and counting unit cubes.

Name _____ Date _____

1. The following solids are made up of 1 cm cubes. Find the total volume of each figure, and write it in the chart below.

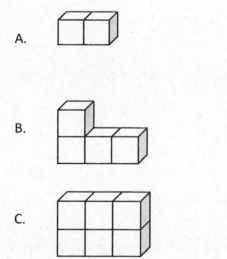

A.

B.

C.

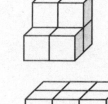

D.

E.

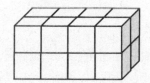

F.

Figure	Volume	Explanation
A		
B		
C		
D		
E		
F		

2. Draw a figure with the given volume on the dot paper.

       a.   3 cubic units   b.   6 cubic units   c.   12 cubic units

3. John built and drew a structure that has a volume of 5 cubic centimeters. His little brother tells him he made a mistake because he only drew 4 cubes. Help John explain to his brother why his drawing is accurate.

4. Draw another figure below that represents a structure with a volume of 5 cubic centimeters.

     Lesson 1:   Explore volume by building with and counting unit cubes.

EUREKA MATH

centimeter grid paper

Lesson 1:   Explore volume by building with and counting unit cubes.

5

©2015 Great Minds. eureka-math.org
G5-M5-SE-BK3-1.3.1-02.2016

This page intentionally left blank

isometric dot paper

This page intentionally left blank

Name _____ Date _____

1. Shade the following figures on centimeter grid paper. Cut and fold each to make 3 open boxes, taping them so they hold their shapes. Pack each box with cubes. Write how many cubes fill each box.

   a.

   Number of cubes: _____

   b.

   Number of cubes: _____

   c.

   Number of cubes: _____

2. Predict how many centimeter cubes will fit in each box, and briefly explain your predictions. Use cubes to find the actual volume. (The figures are not drawn to scale.)

   a.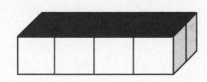

   Prediction: _____

   Actual: _____

Lesson 2: Find the volume of a right rectangular prism by packing with cubic units and counting.

©2015 Great Minds. eureka-math.org
G5-M5-SE-BK3-1.3.1-02.2016

9

b.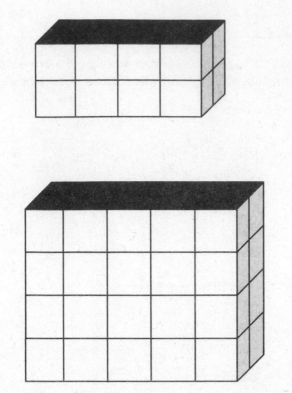

Prediction: _____

Actual: _____

c.

Prediction: _____

Actual: _____

3. Cut out the net in the template, and fold it into a cube. Predict the number of 1-centimeter cubes that would be required to fill it.

a. Prediction: _____

b. Explain your thought process as you made your prediction.

c. How many 1-centimeter cubes are used to fill the figure? Was your prediction accurate?

Lesson 2: Find the volume of a right rectangular prism by packing with cubic units and counting.

 EUREKA MATH

Name _____   Date _____

1. Make the following boxes on centimeter grid paper. Cut and fold each to make 3 open boxes, taping them so they hold their shapes. How many cubes would fill each box? Explain how you found the number.

a. 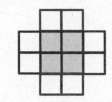   Number of cubes: _____

b. 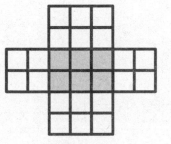   Number of cubes: _____

c. 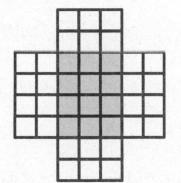   Number of cubes: _____

  EUREKA MATH

Lesson 2:   Find the volume of a right rectangular prism by packing with cubic units and counting.

©2015 Great Minds. eureka-math.org
G5-M5-SE-BK3-1.3.1-02.2016

11

2. How many centimeter cubes would fit inside each box? Explain your answer using words and diagrams on each box. (The figures are not drawn to scale.)

a.

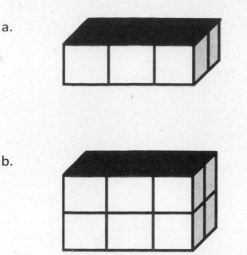

Number of cubes: _____

Explanation:

b.

Number of cubes: _____

Explanation:

c.

Number of cubes: _____

Explanation:

3. The box pattern below holds 24 1-centimeter cubes. Draw two different box patterns that would hold the same number of cubes.

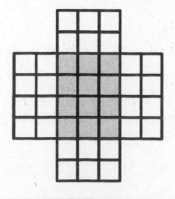

      Lesson 2:   Find the volume of a right rectangular prism by packing with cubic units and counting.

centimeter grid paper (from Lesson 1)

Lesson 2:   Find the volume of a right rectangular prism by packing with cubic
             units and counting.

©2015 Great Minds. eureka-math.org
G5-M5-SE-BK3-1.3.1-02.2016

13

This page intentionally left blank

centimeter grid paper (from Lesson 1)

Lesson 2:   Find the volume of a right rectangular prism by packing with cubic units and counting.

©2015 Great Minds. eureka-math.org
G5-M5-SE-BK3-1.3.1-02.2016

15

This page intentionally left blank

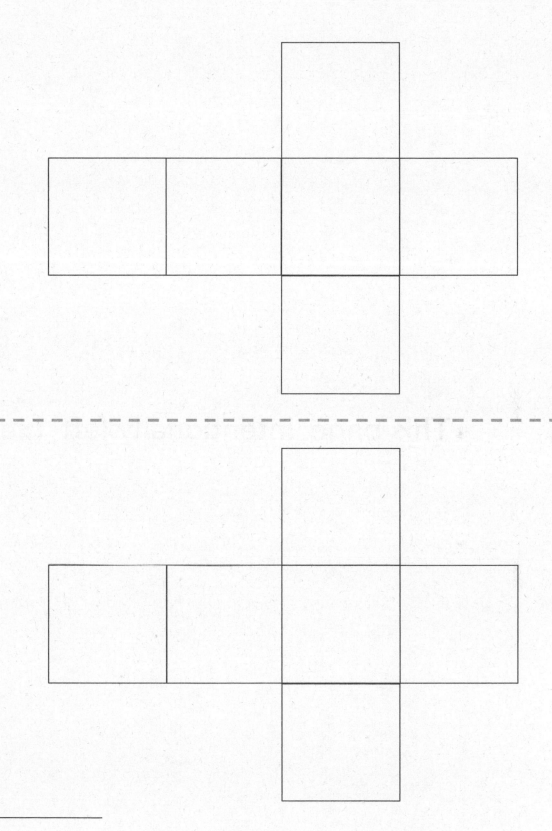

net

Lesson 2:   Find the volume of a right rectangular prism by packing with cubic units and counting.

17

©2015 Great Minds. eureka-math.org
G5-M5-SE-BK3-1.3.1-02.2016

This page intentionally left blank

Name _____ Date _____

1. Use the prisms to find the volume.

- Build the rectangular prism pictured below to the left with your cubes, if necessary.
- Decompose it into layers in three different ways, and show your thinking on the blank prisms.
- Complete the missing information in the table.

a.

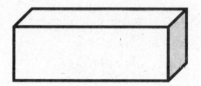

Number of Layers	Number of Cubes in Each Layer	Volume of the Prism
		cubic cm
		cubic cm
		cubic cm

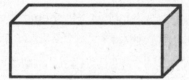
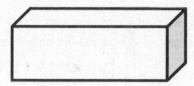

b.

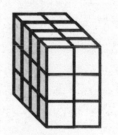

Number of Layers	Number of Cubes in Each Layer	Volume of the Prism
		cubic cm
		cubic cm
		cubic cm

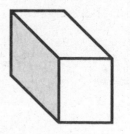
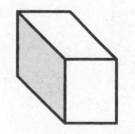
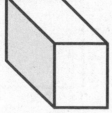

2. Josh and Jonah were finding the volume of the prism to the right. The boys agree that 4 layers can be added together to find the volume. Josh says that he can see on the end of the prism that each layer will have 16 cubes in it. Jonah says that each layer has 24 cubes in it. Who is right? Explain how you know using words, numbers, and/or pictures.

3. Marcos makes a prism 1 inch by 5 inches by 5 inches. He then decides to create layers equal to his first one. Fill in the chart below, and explain how you know the volume of each new prism.

Number of Layers	Volume	Explanation
2		
4		
7		

4. Imagine the rectangular prism below is 6 meters long, 4 meters tall, and 2 meters wide. Draw horizontal lines to show how the prism could be decomposed into layers that are 1 meter in height.

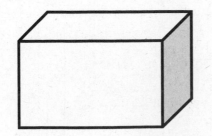

It has _____ layers from bottom to top.

Each horizontal layer contains _____ cubic meters.

The volume of this prism is _____.

Name _____ Date _____

1. Use the prisms to find the volume.

   - The rectangular prisms pictured below were constructed with 1 cm cubes.
   - Decompose each prism into layers in three different ways, and show your thinking on the blank prisms.
   - Complete each table.

a.

Number of Layers	Number of Cubes in Each Layer	Volume of the Prism
		cubic cm
		cubic cm
		cubic cm

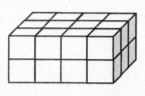

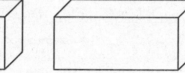
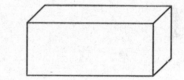
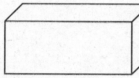

b.

Number of Layers	Number of Cubes in Each Layer	Volume of the Prism
		cubic cm
		cubic cm
		cubic cm

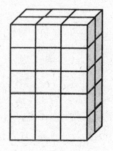

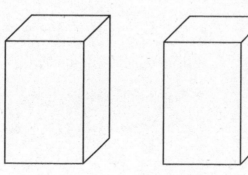
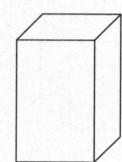

EUREKA MATH™

Lesson 3: Compose and decompose right rectangular prisms using layers.

21

©2015 Great Minds. eureka-math.org
G5-M5-SE-BK3-1.3.1-02.2016

2. Stephen and Chelsea want to increase the volume of this prism by 72 cubic centimeters. Chelsea wants to add eight layers, and Stephen says they only need to add four layers. Their teacher tells them they are both correct. Explain how this is possible.

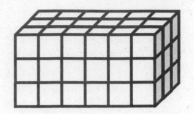

3. Juliana makes a prism 4 inches across and 4 inches wide but only 1 inch tall. She then decides to create layers equal to her first one. Fill in the chart below, and explain how you know the volume of each new prism.

Number of Layers	Volume	Explanation
3		
5		
7		

4. Imagine the rectangular prism below is 4 meters long, 3 meters tall, and 2 meters wide. Draw horizontal lines to show how the prism could be decomposed into layers that are 1 meter in height.

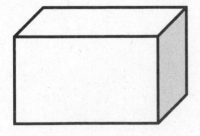

It has _____ layers from top to bottom.

Each horizontal layer contains _____ cubic meters.

The volume of this prism is _____

Name _____   Date _____

Use these rectangular prisms to record the layers that you count.

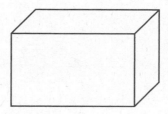 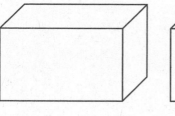 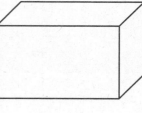 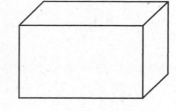

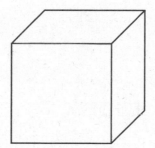 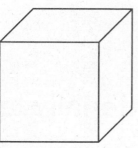 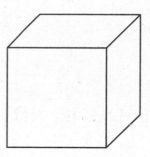 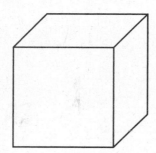

_____

rectangular prism recording sheet

EUREKA MATH

Lesson 3:   Compose and decompose right rectangular prisms using layers.

23

©2015 Great Minds. eureka-math.org
G5-M5-SE-BK3-1.3.1-02.2016

This page intentionally left blank

Name _____   Date _____

1. Each rectangular prism is built from centimeter cubes. State the dimensions, and find the volume.

   a.

   Length: _____ cm

   Width: _____ cm

   Height: _____ cm

   Volume: _____ cm³

   b.

   Length: _____ cm

   Width: _____ cm

   Height: _____ cm

   Volume: _____ cm³

   c.

   Length: _____ cm

   Width: _____ cm

   Height: _____ cm

   Volume: _____ cm³

   d.

   Length: _____ cm

   Width: _____ cm

   Height: _____ cm

   Volume: _____ cm³

2. Write a multiplication sentence that you could use to calculate the volume for each rectangular prism in Problem 1. Include the units in your sentences.

   a. _____   b. _____

   c. _____   d. _____

3. Calculate the volume of each rectangular prism. Include the units in your number sentences.

a.

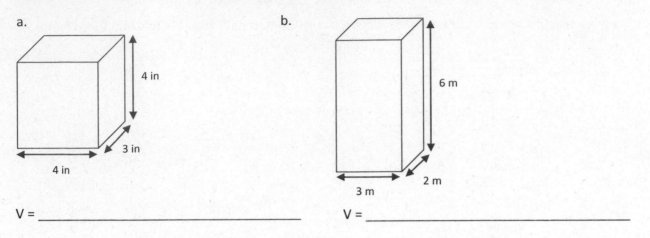

4 in

3 in

4 in

V = _____

b.

6 m

3 m   2 m

V = _____

4. Tyron is constructing a box in the shape of a rectangular prism to store his baseball cards. It has a length of 10 centimeters, a width of 7 centimeters, and a height of 8 centimeters. What is the volume of the box?

5. Aaron says more information is needed to find the volume of the prisms. Explain why Aaron is mistaken, and calculate the volume of the prisms.

a.

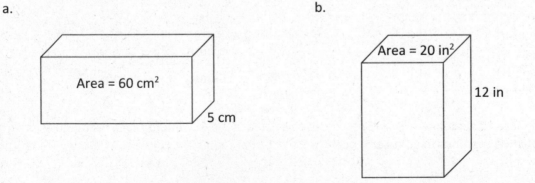

Area = 60 cm²

5 cm

b.

Area = 20 in²

12 in

Lesson 4:   Use multiplication to calculate volume.

©2015 Great Minds. eureka-math.org
G5-M5-SE-BK3-1.3.1-02.2016

Name _____   Date _____

1. Each rectangular prism is built from centimeter cubes. State the dimensions, and find the volume.

a.

Length: _____ cm

Width: _____ cm

Height: _____ cm

Volume: _____ cm³

b.

Length: _____ cm

Width: _____ cm

Height: _____ cm

Volume: _____ cm³

c.

Length: _____ cm

Width: _____ cm

Height: _____ cm

Volume: _____ cm³

d.

Length: _____ cm

Width: _____ cm

Height: _____ cm

Volume: _____ cm³

2. Write a multiplication sentence that you could use to calculate the volume for each rectangular prism in Problem 1. Include the units in your sentences.

a. _____   b. _____

c. _____   d. _____

3. Calculate the volume of each rectangular prism. Include the units in your number sentences.

a.

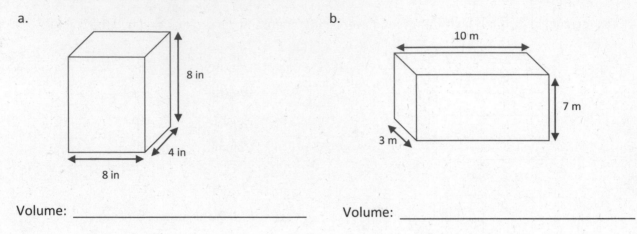

8 in

4 in

8 in

b.

10 m

7 m

3 m

Volume: _____

Volume: _____

4. Mrs. Johnson is constructing a box in the shape of a rectangular prism to store clothes for the summer. It has a length of 28 inches, a width of 24 inches, and a height of 30 inches. What is the volume of the box?

5. Calculate the volume of each rectangular prism using the information that is provided.

a. Face area: 56 square meters

Height: 4 meters

b. Face area: 169 square inches

Height: 14 inches

Lesson 4:   Use multiplication to calculate volume.

©2015 Great Minds. eureka-math.org
G5-M5-SE-BK3-1.3.1-02.2016

Name _____   Date _____

Use these rectangular prisms to record the layers that you count.

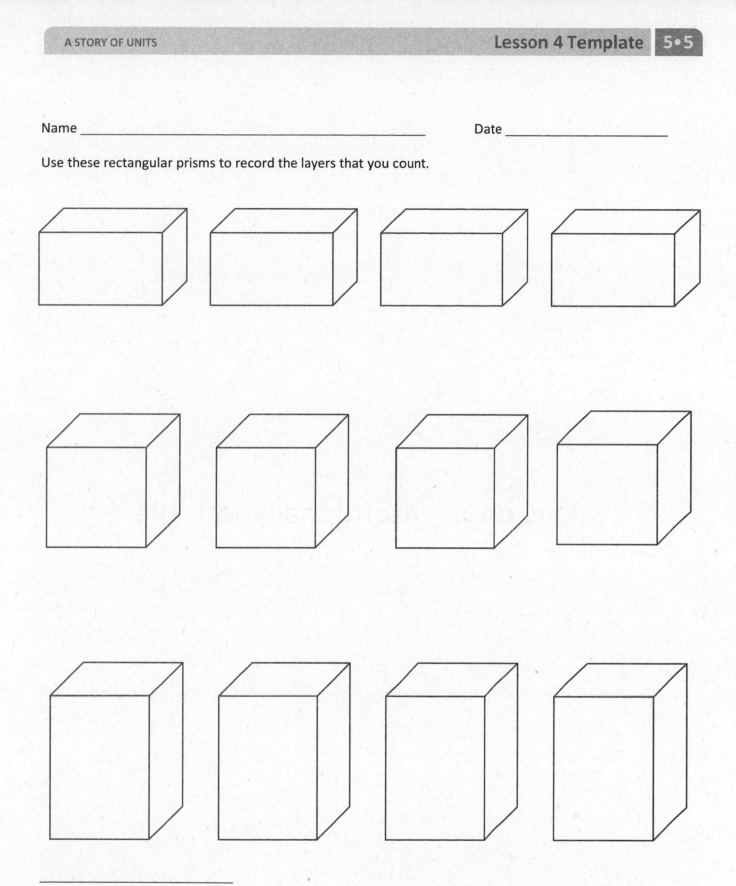

_____

rectangular prism recording sheet (from Lesson 3)

This page intentionally left blank

Name _____ Date _____

1. Determine the volume of two boxes on the table using cubes, and then confirm by measuring and multiplying.

Box Number	Number of Cubes Packed	Measurements			Volume
		Length	Width	Height	

2. Using the same boxes from Problem 1, record the amount of liquid that your box can hold.

Box Number	Liquid the Box Can Hold
	mL
	mL

3. Shade to show the water in the graduated cylinder.

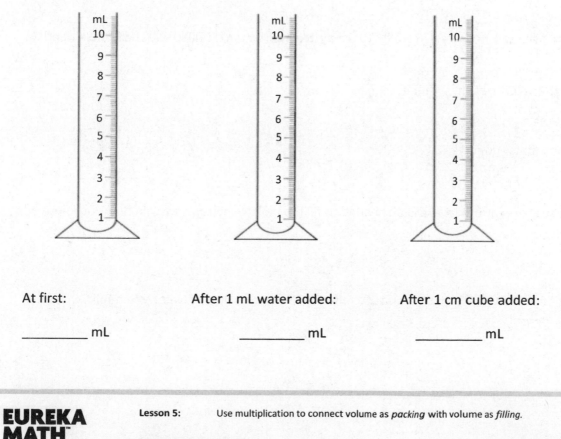

At first: After 1 mL water added: After 1 cm cube added:

_____ mL   _____ mL   _____ mL

EUREKA
MATH

Lesson 5:   Use multiplication to connect volume as *packing* with volume as *filling*.

31

©2015 Great Minds. eureka-math.org
G5-M5-SE-BK3-1.3.1-02.2016

4. What conclusion can you draw about 1 cubic centimeter and 1 mL?

5. The tank, shaped like a rectangular prism, is filled to the top with water.

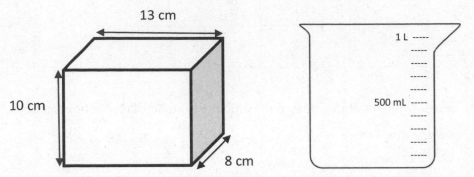

Will the beaker hold all the water in the tank? If yes, how much more will the beaker hold?
If no, how much more will the tank hold than the beaker? Explain how you know.

6. A rectangular fish tank measures 26 cm by 20 cm by 18 cm. The tank is filled with water to a depth of 15 cm.

   a. What is the volume of the water in mL?

   b. How many liters is that?

   c. How many more mL of water will be needed to fill the tank to the top? Explain how you know.

7. A rectangular container is 25 cm long and 20 cm wide. If it holds 1 liter of water when full, what is its height?

©2015 Great Minds. eureka-math.org
G5-M5-SE-BK3-1.3.1-02.2016

Name _____   Date _____

1. Johnny filled a container with 30 centimeter cubes. Shade the beaker to show how much water the container will hold. Explain how you know.

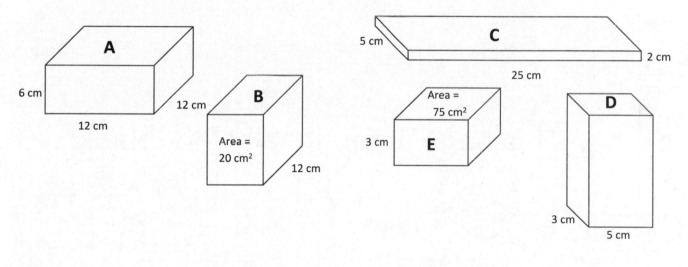

2. A beaker contains 250 mL of water. Jack wants to pour the water into a container that will hold the water. Which of the containers pictured below could he use? Explain your choices.

3. On the back of this paper, describe the details of the activities you did in class today. Include what you learned about cubic centimeters and milliliters. Give an example of a problem you solved with an illustration.

Lesson 5:   Use multiplication to connect volume as *packing* with volume as *filling*.

©2015 Great Minds. eureka-math.org
G5-M5-SE-BK3-1.3.1-02.2016

33

This page intentionally left blank

Name _____ Date _____

1. Find the total volume of the figures, and record your solution strategy.

   a.

   5 cm

   5 cm

   3 cm

   14 cm

   Volume: _____

   Solution Strategy:

   b.

   7 in

   3 in

   6 in

   4 in

   15 in

   Volume: _____

   Solution Strategy:

   c.

   4 cm

   6 cm

   2 cm

   3 cm

   10 cm

   Volume: _____

   Solution Strategy:

   d.

   8 m

   12 m

   6 m

   3 m

   10 m

   Volume: _____

   Solution Strategy:

Lesson 6:   Find the total volume of solid figures composed of two non-overlapping rectangular prisms.

35

©2015 Great Minds. eureka-math.org
G5-M5-SE-BK3-1.3.1-02.2016

2. A sculpture (pictured below) is made of two sizes of rectangular prisms. One size measures 13 in by 8 in by 2 in. The other size measures 9 in by 8 in by 18 in. What is the total volume of the sculpture?

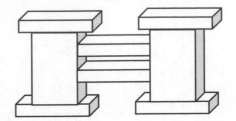

3. The combined volume of two identical cubes is 128 cubic centimeters. What is the side length of each cube?

4. A rectangular tank with a base area of 24 cm² is filled with water and oil to a depth of 9 cm. The oil and water separate into two layers when the oil rises to the top. If the thickness of the oil layer is 4 cm, what is the volume of the water?

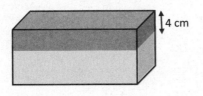

4 cm

5. Two rectangular prisms have a combined volume of 432 cubic feet. Prism A has half the volume of Prism B.

   a. What is the volume of Prism A? Prism B?

   b. If Prism A has a base area of 24 ft², what is the height of Prism A?

   c. If Prism B's base is $\frac{2}{3}$ the area of Prism A's base, what is the height of Prism B?

Lesson 6:   Find the total volume of solid figures composed of two non-overlapping rectangular prisms.

©2015 Great Minds. eureka-math.org
G5-M5-SE-BK3-1.3.1-02.2016

EUREKA MATH

Name _____   Date _____

1. Find the total volume of the figures, and record your solution strategy.

    a.   b.

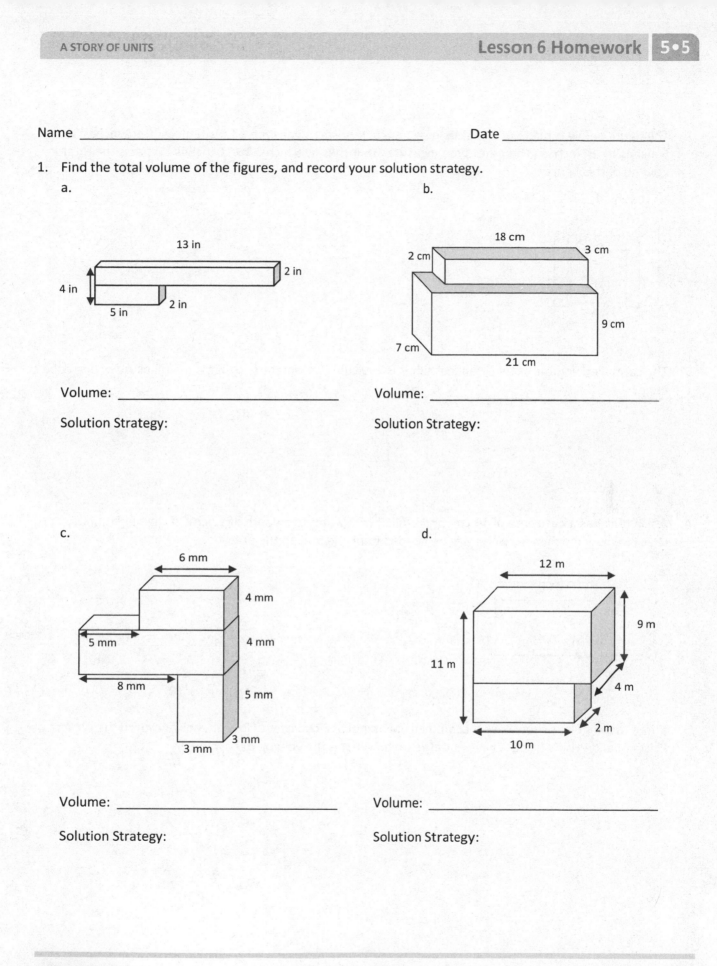

    Volume: _____   Volume: _____

    Solution Strategy: Solution Strategy:

    c.   d.

    Volume: _____   Volume: _____

    Solution Strategy: Solution Strategy:

2. The figure below is made of two sizes of rectangular prisms. One type of prism measures 3 inches by 6 inches by 14 inches. The other type measures 15 inches by 5 inches by 10 inches. What is the total volume of this figure?

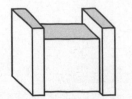

3. The combined volume of two identical cubes is 250 cubic centimeters. What is the measure of one cube's edge?

4. A fish tank has a base area of 45 cm² and is filled with water to a depth of 12 cm. If the height of the tank is 25 cm, how much more water will be needed to fill the tank to the brim?

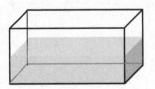

5. Three rectangular prisms have a combined volume of 518 cubic feet. Prism A has one-third the volume of Prism B, and Prisms B and C have equal volume. What is the volume of each prism?

Lesson 6:   Find the total volume of solid figures composed of two non-overlapping rectangular prisms.

Name _____ Date _____

Geoffrey builds rectangular planters.

1. Geoffrey's first planter is 8 feet long and 2 feet wide. The container is filled with soil to a height of 3 feet in the planter. What is the volume of soil in the planter? Explain your work using a diagram.

2. Geoffrey wants to grow some tomatoes in four large planters. He wants each planter to have a volume of 320 cubic feet, but he wants them all to be different. Show four different ways Geoffrey can make these planters, and draw diagrams with the planters' measurements on them.

Planter A	Planter B
Planter C	Planter D

Lesson 7: Solve word problems involving the volume of rectangular prisms with whole number edge lengths.

39

©2015 Great Minds. eureka-math.org
G5-M5-SE-BK3-1.3.1-02.2016

3. Geoffrey wants to make one planter that extends from the ground to just below his back window.
    The window starts 3 feet off the ground. If he wants the planter to hold 36 cubic feet of soil, name one
    way he could build the planter so it is not taller than 3 feet. Explain how you know.

4. After all of this gardening work, Geoffrey decides he needs a new shed to replace the old one. His current
    shed is a rectangular prism that measures 6 feet long by 5 feet wide by 8 feet high. He realizes he needs a
    shed with 480 cubic feet of storage.

    a. Will he achieve his goal if he doubles each dimension? Why or why not?

    b. If he wants to keep the height the same, what could the other dimensions be for him to get the
        volume he wants?

    c. If he uses the dimensions in part (b), what could be the area of the new shed's floor?

Lesson 7:   Solve word problems involving the volume of rectangular prisms with
              whole number edge lengths.

©2015 Great Minds. eureka-math.org
G5-M5-SE-BK3-1.3.1-02.2016

Name _____ Date _____

Wren makes some rectangular display boxes.

1. Wren's first display box is 6 inches long, 9 inches wide, and 4 inches high. What is the volume of the display box? Explain your work using a diagram.

2. Wren wants to put some artwork into three shadow boxes. She knows they all need a volume of 60 cubic inches, but she wants them all to be different. Show three different ways Wren can make these boxes by drawing diagrams and labeling the measurements.

Shadow Box A	Shadow Box B
Shadow Box C	

Lesson 7: Solve word problems involving the volume of rectangular prisms with whole number edge lengths.

©2015 Great Minds. eureka-math.org
G5-M5-SE-BK3-1.3.1-02.2016

41

3. Wren wants to build a box to organize her scrapbook supplies. She has a stencil set that is 12 inches wide that needs to lay flat in the bottom of the box. The supply box must also be no taller than 2 inches. Name one way she could build a supply box with a volume of 72 cubic inches.

4. After all of this organizing, Wren decides she also needs more storage for her soccer equipment. Her current storage box measures 1 foot long by 2 feet wide by 2 feet high. She realizes she needs to replace it with a box with 12 cubic feet of storage, so she doubles the width.

    a. Will she achieve her goal if she does this? Why or why not?

    b. If she wants to keep the height the same, what could the other dimensions be for a 12-cubic-foot storage box?

    c. If she uses the dimensions in part (b), what is the area of the new storage box's floor?

    d. How has the area of the bottom in her new storage box changed? Explain how you know.

Lesson 7:   Solve word problems involving the volume of rectangular prisms with
                          whole number edge lengths.

©2015 Great Minds. eureka-math.org
G5-M5-SE-BK3-1.3.1-02.2016

Name _____   Date _____

Using the box patterns, construct a sculpture containing at least 5, but not more than 7, rectangular prisms that meets the following requirements in the table below.

1.	My sculpture has 5 to 7 rectangular prisms.	Number of prisms: _____
2.	Each prism is labeled with a letter, dimensions, and volume.	

Prism A _____ by _____ by _____   Volume = _____

Prism B _____ by _____ by _____   Volume = _____

Prism C _____ by _____ by _____   Volume = _____

Prism D _____ by _____ by _____   Volume = _____

Prism E _____ by _____ by _____   Volume = _____

Prism __ _____ by _____ by _____   Volume = _____

Prism __ _____ by _____ by _____   Volume = _____

3.	Prism D has $\frac{1}{2}$ the volume of Prism ____.	Prism D Volume = _____ Prism ____ Volume = _____
4.	Prism E has $\frac{1}{3}$ the volume of Prism ____.	Prism E Volume = _____ Prism ____ Volume = _____
5.	The total volume of all the prisms is 1,000 cubic centimeters or less.	Total volume: _____ Show calculations:

Lesson 8:   Apply concepts and formulas of volume to design a sculpture using rectangular prisms within given parameters.

43

©2015 Great Minds. eureka-math.org
G5-M5-SE-BK3-1.3.1-02.2016

This page intentionally left blank

Name _____   Date _____

1. I have a prism with the dimensions of 6 cm by 12 cm by 15 cm. Calculate the volume of the prism, and then give the dimensions of three different prisms that each have $\frac{1}{3}$ of the volume.

	Length	Width	Height	Volume
Original Prism	6 cm	12 cm	15 cm	
Prism 1				
Prism 2				
Prism 3				

2. Sunni's bedroom has the dimensions of 11 ft by 10 ft by 10 ft. Her den has the same height but double the volume. Give two sets of the possible dimensions of the den and the volume of the den.

Lesson 8:   Apply concepts and formulas of volume to design a sculpture using   45
              rectangular prisms within given parameters.

©2015 Great Minds. eureka-math.org
G5-M5-SE-BK3-1.3.1-02.2016

This page intentionally left blank

Project Requirements

1. Each project must include 5 to 7 rectangular prisms.
2. All prisms must be labeled with a letter (beginning with A), dimensions, and volume.
3. Prism D must be $\frac{1}{2}$ the volume of another prism.
4. Prism E must be $\frac{1}{3}$ the volume of another prism.
5. The total volume of all of the prisms must be 1,000 cubic centimeters or less.

Project Requirements

1. Each project must include 5 to 7 rectangular prisms.
2. All prisms must be labeled with a letter (beginning with A), dimensions, and volume.
3. Prism D must be $\frac{1}{2}$ the volume of another prism.
4. Prism E must be $\frac{1}{3}$ the volume of another prism.
5. The total volume of all of the prisms must be 1,000 cubic centimeters or less.

Project Requirements

1. Each project must include 5 to 7 rectangular prisms.
2. All prisms must be labeled with a letter (beginning with A), dimensions, and volume.
3. Prism D must be $\frac{1}{2}$ the volume of another prism.
4. Prism E must be $\frac{1}{3}$ the volume of another prism.
5. The total volume of all of the prisms must be 1,000 cubic centimeters or less.

---

project requirements

Lesson 8:   Apply concepts and formulas of volume to design a sculpture using
rectangular prisms within given parameters.

47

This page intentionally left blank

Note: Be sure to set printer to *actual size* before printing.

box pattern (a)

Lesson 8: Apply concepts and formulas of volume to design a sculpture using rectangular prisms within given parameters.

©2015 Great Minds. eureka-math.org
G5-M5-SE-BK3-1.3.1-02.2016

49

This page intentionally left blank

box pattern (b)

EUREKA MATH

Lesson 8: Apply concepts and formulas of volume to design a sculpture using rectangular prisms within given parameters.

51

This page intentionally left blank

box pattern (c)

Lesson 8:   Apply concepts and formulas of volume to design a sculpture using
rectangular prisms within given parameters.

53

©2015 Great Minds. eureka-math.org
G5-M5-SE-BK3-1.3.1-02.2016

This page intentionally left blank

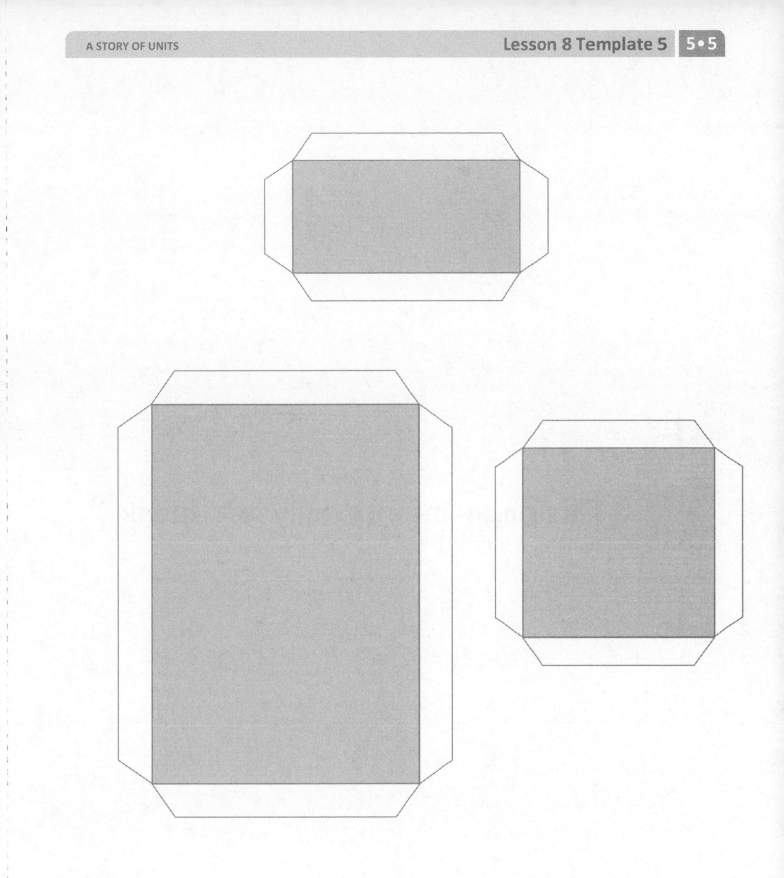

lid patterns

EUREKA
MATH™

Lesson 8:   Apply concepts and formulas of volume to design a sculpture using
rectangular prisms within given parameters.

55

©2015 Great Minds. eureka-math.org
G5-M5-SE-BK3-1.3.1-02.2016

This page intentionally left blank

Name _____   Date _____

Evaluation Rubric

CATEGORY	4	3	2	1	Subtotal
Completeness of Personal Project and Classmate Evaluation	All components of the project are present and correct, and a detailed evaluation of a classmate's project has been completed.	Project is missing 1 component, and a detailed evaluation of a classmate's project has been completed.	Project is missing 2 components, and an evaluation of a classmate's project has been completed.	Project is missing 3 or more components, and an evaluation of a classmate's project has been completed.	(× 4) _____/16
Accuracy of Calculations	Volume calculations for all prisms are correct.	Volume calculations include 1 error.	Volume calculations include 2–3 errors.	Volume calculations include 4 or more errors.	(× 5) _____/20
Neatness and Use of Color	All elements of the project are carefully and colorfully constructed.	Some elements of the project are carefully and colorfully constructed.	Project lacks color or is not carefully constructed.	Project lacks color and is not carefully constructed.	(× 2) _____/4
					TOTAL: _____/40

_____

evaluation rubric

Lesson 8:   Apply concepts and formulas of volume to design a sculpture using rectangular prisms within given parameters.

57

©2015 Great Minds. eureka-math.org
G5-M5-SE-BK3-1.3.1-02.2016

This page intentionally left blank

Name _____ Date _____

I reviewed project number _____.

Use the rubric below to evaluate your friend's project. Ask questions and measure the parts to determine whether your friend has all the required elements. Respond to the prompt in italics in the third column. The final column can be used to write something you find interesting about that element if you like.

Space is provided beneath the rubric for your calculations.

	Requirement	Element Present? (✓)	Specifics of Element	Notes
1.	The sculpture has 5 to 7 prisms.		*# of prisms:*	
2.	All prisms are labeled with a letter.		*Write letters used:*	
3.	All prisms have correct dimensions with units written on the top.		*List any prisms with incorrect dimensions or units:*	
4.	All prisms have correct volume with units written on the top.		*List any prism with incorrect dimensions or units:*	
5.	Prism D has $\frac{1}{2}$ the volume of another prism.		*Record on next page:*	
6.	Prism E has $\frac{1}{3}$ the volume of another prism.		*Record on next page:*	
7.	The total volume of all the parts together is 1,000 cubic units or less.		*Total volume:*	

Calculations:

Lesson 9: Apply concepts and formulas of volume to design a sculpture using rectangular prisms within given parameters.

59

©2015 Great Minds. eureka-math.org
G5-M5-SE-BK3-1.3.1-02.2016

8. Measure the dimensions of each prism. Calculate the volume of each prism and the total volume. Record that information in the table below. If your measurements or volume differ from those listed on the project, put a star by the prism label in the table below, and record on the rubric.

Prism	Dimensions	Volume
A	_____ by _____ by _____	
B	_____ by _____ by _____	
C	_____ by _____ by _____	
D	_____ by _____ by _____	
E	_____ by _____ by _____	
	_____ by _____ by _____	
	_____ by _____ by _____	

9. Prism D's volume is $\frac{1}{2}$ that of Prism _____.
   Show calculations below.

10. Prism E's volume is $\frac{1}{3}$ that of Prism _____.
    Show calculations below.

11. Total volume of sculpture: _____.
    Show calculations below.

Lesson 9:   Apply concepts and formulas of volume to design a sculpture using
               rectangular prisms within given parameters.

©2015 Great Minds. eureka-math.org
G5-M5-SE-BK3-1.3.1-02.2016

Name _____   Date _____

1. Find three rectangular prisms around your house. Describe the item you are measuring (cereal box, tissue box, etc.), and then measure each dimension to the nearest whole inch, and calculate the volume.

   a. Rectangular Prism A

      Item:

      Height: _____ inches

      Length: _____ inches

      Width: _____ inches

      Volume: _____ cubic inches

   b. Rectangular Prism B

      Item:

      Height: _____ inches

      Length: _____ inches

      Width: _____ inches

      Volume: _____ cubic inches

   c. Rectangular Prism C

      Item:

      Height: _____ inches

      Length: _____ inches

      Width: _____ inches

      Volume: _____ cubic inches

Lesson 9:   Apply concepts and formulas of volume to design a sculpture using
              rectangular prisms within given parameters.

©2015 Great Minds. eureka-math.org
G5-M5-SE-BK3-1.3.1-02.2016

61

This page intentionally left blank

Name _____   Date _____

Sketch the rectangles and your tiling. Write the dimensions and the units you counted in the blanks.
Then, use multiplication to confirm the area. Show your work. We will do Rectangles A and B together.

1. **Rectangle A:**   Rectangle A is

                                                  _____ units long _____ units wide

                                                  Area = _____ units²

2. **Rectangle B:**   3. **Rectangle C:**

Rectangle B is   Rectangle C is

_____ units long _____ units wide   _____ units long _____ units wide

Area = _____ units²   Area = _____ units²

4. **Rectangle D:**   5. **Rectangle E:**

Rectangle D is   Rectangle E is

_____ units long _____ units wide   _____ units long _____ units wide

Area = _____ units²   Area = _____ units²

EUREKA MATH™   **Lesson 10:**   Find the area of rectangles with whole-by-mixed and
whole-by-fractional number side lengths by tiling, record by drawing,
and relate to fraction multiplication.   63

©2015 Great Minds. eureka-math.org
G5-M5-SE-BK3-1.3.1-02.2016

6. The rectangle to the right is composed of squares that measure $2\frac{1}{4}$ inches on each side. What is its area in square inches? Explain your thinking using pictures and numbers.

7. A rectangle has a perimeter of $35\frac{1}{2}$ feet. If the length is 12 feet, what is the area of the rectangle?

Lesson 10: Find the area of rectangles with whole-by-mixed and whole-by-fractional number side lengths by tiling, record by drawing, and relate to fraction multiplication.

Name _____ Date _____

1. John tiled some rectangles using square units. Sketch the rectangles if necessary. Fill in the missing information, and then confirm the area by multiplying.

   a. **Rectangle A:**

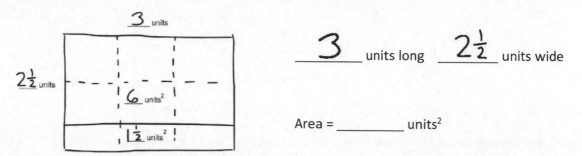

   Rectangle A is

   ___3___ units long   ___$2\frac{1}{2}$___ units wide

   Area = _____ units²

   b. **Rectangle B:**

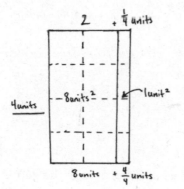

   Rectangle B is

   _____ units long   _____ units wide

   Area = _____ units²

   c. **Rectangle C:**

   Rectangle C is

   ___$\frac{3}{4}$___ units long   ___4___ units wide

   Area = _____ units²

Lesson 10:   Find the area of rectangles with whole by mixed and whole-by-fractional number side lengths by tiling, record by drawing, and relate to fraction multiplication.

©2015 Great Minds. eureka-math.org
G5-M5-SE-BK3-1.3.1-02.2016

65

d. **Rectangle D:**

Rectangle D is

_____2_____ units long _____1¾_____ units wide

Area = _____ units²

2. Rachel made a mosaic from different color rectangular tiles. Three tiles measured $3\frac{1}{2}$ inches × 3 inches. Six tiles measured 4 inches × $3\frac{1}{4}$ inches. What is the area of the whole mosaic in square inches?

3. A garden box has a perimeter of $27\frac{1}{2}$ feet. If the length is 9 feet, what is the area of the garden box?

Lesson 10: Find the area of rectangles with whole-by-mixed and whole-by-fractional number side lengths by tiling, record by drawing, and relate to fraction multiplication.

Name _____   Date _____

Draw the rectangle and your tiling.
Write the dimensions and the units you counted in the blanks.
Then, use multiplication to confirm the area. Show your work.

1. **Rectangle A:**

2. **Rectangle B:**

Rectangle A is

_____ units long   _____ units wide

Area = _____ units^2

Rectangle B is

_____ units long   _____ units wide

Area = _____ units^2

3. **Rectangle C:**

4. **Rectangle D:**

Rectangle C is

_____ units long   _____ units wide

Area = _____ units^2

Rectangle D is

_____ units long   _____ units wide

Area = _____ units^2

Lesson 11:   Find the area of rectangles with mixed-by-mixed and fraction-by-fraction side lengths by tiling, record by drawing, and relate to fraction multiplication.

67

EUREKA
MATH™

5. Colleen and Caroline each built a rectangle out of square tiles placed in 3 rows of 5. Colleen used tiles that measured $1\frac{2}{3}$ cm in length. Caroline used tiles that measured $3\frac{1}{3}$ cm in length.

    a. Draw the girls' rectangles, and label the lengths and widths of each.

    b. What are the areas of the rectangles in square centimeters?

    c. Compare the areas of the rectangles.

6. A square has a perimeter of 51 inches. What is the area of the square?

Lesson 11: Find the area of rectangles with mixed-by-mixed and fraction-by-fraction side lengths by tiling, record by drawing, and relate to fraction multiplication.

Name _____   Date _____

1. Kristen tiled the following rectangles using square units. Sketch the rectangles, and find the areas. Then, confirm the area by multiplying. Rectangle A has been sketched for you.

a. **Rectangle A:**

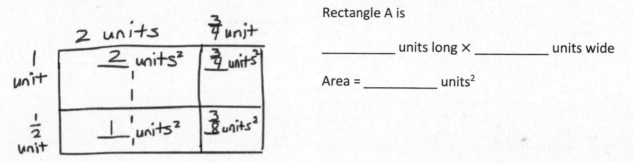

Rectangle A is

_____ units long × _____ units wide

Area = _____ units²

b. **Rectangle B:**

Rectangle B is

$2\frac{1}{2}$ units long × $\frac{3}{4}$ unit wide

Area = _____ units²

c. **Rectangle C:**

Rectangle C is

$3\frac{1}{3}$ units long × $2\frac{1}{2}$ units wide

Area = _____ units²

d.   **Rectangle D:**

Rectangle D is

$3\frac{1}{2}$ units long $\times\ 2\frac{1}{4}$ units wide

Area = _____ units^2

2.   A square has a perimeter of 25 inches. What is the area of the square?

Lesson 11:   Find the area of rectangles with mixed-by-mixed and fraction-by-fraction side lengths by tiling, record by drawing, and relate to fraction multiplication.

©2015 Great Minds. eureka-math.org
G5-M5-SE-BK3-1.3.1-02.2016

Name _____ Date _____

1. Measure each rectangle to the nearest $\frac{1}{4}$ inch with your ruler, and label the dimensions. Use the area model to find each area.

a.

b.

c.

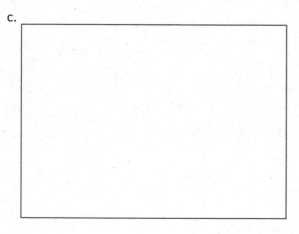

d.

Lesson 12:   Measure to find the area of rectangles with fractional side lengths.

71

©2015 Great Minds. eureka-math.org
G5-M5-SE-BK3-1.3.1-02.2016

e.

f.

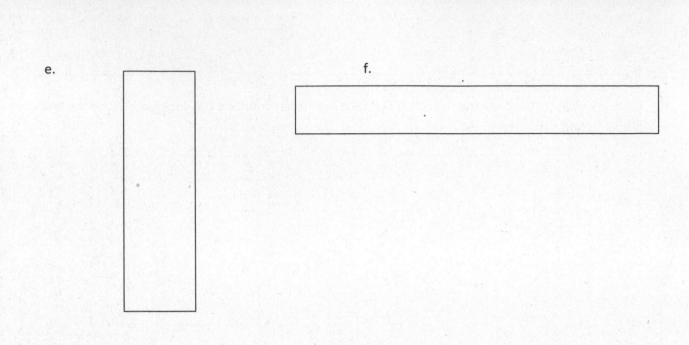

2. Find the area of rectangles with the following dimensions. Explain your thinking using the area model.

a. $1 \text{ ft} \times 1\frac{1}{2} \text{ ft}$

b. $1\frac{1}{2} \text{ yd} \times 1\frac{1}{2} \text{ yd}$

c. $2\frac{1}{2} \text{ yd} \times 1\frac{3}{16} \text{ yd}$

Lesson 12: Measure to find the area of rectangles with fractional side lengths.

3. Hanley is putting carpet in her house. She wants to carpet her living room, which measures 15 ft × $12\frac{1}{3}$ ft. She also wants to carpet her dining room, which is $10\frac{1}{4}$ ft × $10\frac{1}{3}$ ft. How many square feet of carpet will she need to cover both rooms?

4. Fred cut a $9\frac{3}{4}$-inch square of construction paper for an art project. He cut a square from the edge of the big rectangle whose sides measured $3\frac{1}{4}$ inches. (See the picture below.)

    a. What is the area of the smaller square that Fred cut out?

    b. What is the area of the remaining paper?

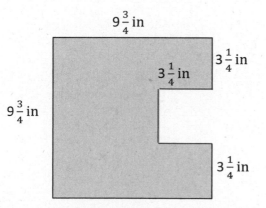

This page intentionally left blank

Name _____ Date _____

1. Measure each rectangle to the nearest $\frac{1}{4}$ inch with your ruler, and label the dimensions. Use the area model to find the area.

a.

b.

c.

d.

e.

Lesson 12: Measure to find the area of rectangles with fractional side lengths.

75

©2015 Great Minds. eureka-math.org
G5-M5-SE-BK3-1.3.1-02.2016

2. Find the area of rectangles with the following dimensions. Explain your thinking using the area model.

   a. $2\frac{1}{4}\,yd \times \frac{1}{4}\,yd$

   b. $2\frac{1}{2}\,ft \times 1\frac{1}{4}\,ft$

3. Kelly buys a tarp to cover the area under her tent. The tent is 4 feet wide and has an area of 31 square feet. The tarp she bought is $5\frac{1}{3}$ feet by $5\frac{3}{4}$ feet. Can the tarp cover the area under Kelly's tent? Draw a model to show your thinking.

4. Shannon and Leslie want to carpet a $16\frac{1}{2}$-ft by $16\frac{1}{2}$-ft square room. They cannot put carpet under an entertainment system that juts out. (See the drawing below.)

   a. In square feet, what is the area of the space with no carpet?

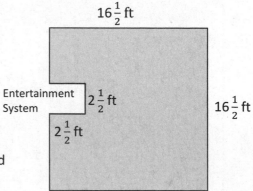

   b. How many square feet of carpet will Shannon and Leslie need to buy?

Lesson 12:   Measure to find the area of rectangles with fractional side lengths.

Name _____ Date _____

1. Find the area of the following rectangles. Draw an area model if it helps you.

   a. $\frac{5}{4}$ km $\times \frac{12}{5}$ km

   b. $16\frac{1}{2}$ m $\times 4\frac{1}{5}$ m

   c. $4\frac{1}{3}$ yd $\times 5\frac{2}{3}$ yd

   d. $\frac{7}{8}$ mi $\times 4\frac{1}{3}$ mi

2. Julie is cutting rectangles out of fabric to make a quilt. If the rectangles are $2\frac{3}{5}$ inches wide and $3\frac{2}{3}$ inches long, what is the area of four such rectangles?

Lesson 13: Multiply mixed number factors, and relate to the distributive property and the area model.

©2015 Great Minds. eureka-math.org
G5-M5-SE-BK3-1.3.1-02.2016

77

3. Mr. Howard's pool is connected to his pool house by a sidewalk as shown. He wants to buy sod for the lawn, shown in gray. How much sod does he need to buy?

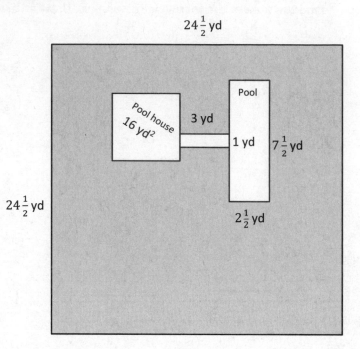

Lesson 13:   Multiply mixed number factors, and relate to the distributive property and the area model.

©2015 Great Minds. eureka-math.org
G5-M5-SE-BK3-1.3.1-02.2016

Name _____ Date _____

1. Find the area of the following rectangles. Draw an area model if it helps you.

a. $\frac{8}{3}$ cm × $\frac{24}{4}$ cm

b. $\frac{32}{5}$ ft × $3\frac{3}{8}$ ft

c. $5\frac{4}{6}$ in × $4\frac{3}{5}$ in

d. $\frac{5}{7}$ m × $6\frac{3}{5}$ m

2. Chris is making a tabletop from some leftover tiles. He has 9 tiles that measure $3\frac{1}{8}$ inches long and $2\frac{3}{4}$ inches wide. What is the greatest area he can cover with these tiles?

Lesson 13: Multiply mixed number factors, and relate to the distributive property and the area model.

79

©2015 Great Minds. eureka-math.org
G5-M5-SE-BK3-1.3.1-02.2016

3. A hotel is recarpeting a section of the lobby. Carpet covers the part of the floor as shown below in gray. How many square feet of carpeting will be needed?

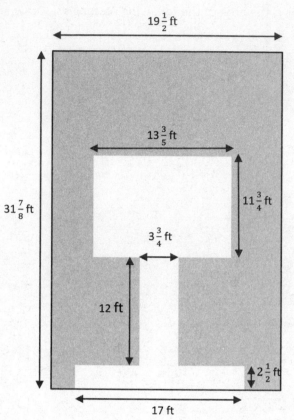

Lesson 13:   Multiply mixed number factors, and relate to the distributive property and the area model.

©2015 Great Minds. eureka-math.org
G5-M5-SE-BK3-1.3.1-02.2016

EUREKA
MATH™

Name _____   Date _____

1. George decided to paint a wall with two windows. Both windows are $3\frac{1}{2}$-ft by $4\frac{1}{2}$-ft rectangles. Find the area the paint needs to cover.

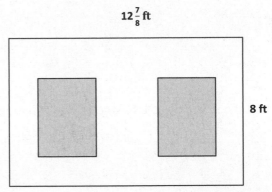

2. Joe uses square tiles, some of which he cuts in half, to make the figure below. If each square tile has a side length of $2\frac{1}{2}$ inches, what is the total area of the figure?

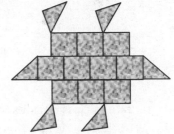

3. All-In-One Carpets is installing carpeting in three rooms. How many square feet of carpet are needed to carpet all three rooms?

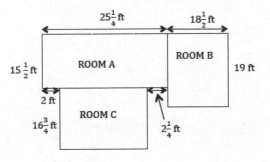

Lesson 14:   Solve real-world problems involving area of figures with fractional side lengths using visual models and/or equations.   81

©2015 Great Minds. eureka-math.org
G5-M5-SE-BK3-1.3.1-02.2016

4. Mr. Johnson needs to buy sod for his front lawn.

   a. If the lawn measures $36\frac{2}{3}$ ft by $45\frac{1}{6}$ ft, how many square feet of sod will he need?

   b. If sod is only sold in whole square feet, how much will Mr. Johnson have to pay?

   Sod Prices

Area	Price per Square Foot
First 1,000 sq ft	$0.27
Next 500 sq ft	$0.22
Additional square feet	$0.19

5. Jennifer's class decides to make a quilt. Each of the 24 students will make a quilt square that is 8 inches on each side. When they sew the quilt together, every edge of each quilt square will lose $\frac{3}{4}$ of an inch.

   a. Draw one way the squares could be arranged to make a rectangular quilt. Then, find the perimeter of your arrangement.

   b. Find the area of the quilt.

Lesson 14: Solve real-world problems involving area of figures with fractional side lengths using visual models and/or equations.

©2015 Great Minds. eureka-math.org
G5-M5-SE-BK3-1.3.1-02.2016

Name _____   Date _____

1. Mr. Albano wants to paint menus on the wall of his café in chalkboard paint. The gray area below shows where the rectangular menus will be. Each menu will measure 6-ft wide and $7\frac{1}{2}$-ft tall.

    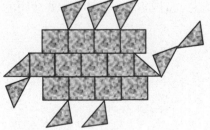

    25 ft

    $13\frac{2}{3}$ ft

    - How many square feet of menu space will Mr. Albano have?

    - What is the area of wall space that is not covered by chalkboard paint?

2. Mr. Albano wants to put tiles in the shape of a dinosaur at the front entrance. He will need to cut some tiles in half to make the figure. If each square tile is $4\frac{1}{4}$ inches on each side, what is the total area of the dinosaur?

Lesson 14:   Solve real-world problems involving area of figures with fractional side lengths using visual models and/or equations.

©2015 Great Minds. eureka-math.org
G5-M5-SE-BK3-1.3.1-02.2016

83

3. A-Plus Glass is making windows for a new house that is being built. The box shows the list of sizes they must make.

15 windows	$4\frac{3}{4}$-ft long and $3\frac{3}{5}$-ft wide
7 windows	$2\frac{4}{5}$-ft wide and $6\frac{1}{2}$-ft long

    How many square feet of glass will they need?

4. Mr. Johnson needs to buy seed for his backyard lawn.

    ▪ If the lawn measures $40\frac{4}{5}$ ft by $50\frac{7}{8}$ ft, how many square feet of seed will he need to cover the entire area?

    ▪ One bag of seed will cover 500 square feet if he sets his seed spreader to its highest setting and 300 square feet if he sets the spreader to its lowest setting. How many bags of seed will he need if he uses the highest setting? The lowest setting?

Lesson 14:   Solve real-world problems involving area of figures with fractional side lengths using visual models and/or equations.

Name _____   Date _____

1. The length of a flowerbed is 4 times as long as its width. If the width is $\frac{3}{8}$ meter, what is the area?

2. Mrs. Johnson grows herbs in square plots. Her basil plot measures $\frac{5}{8}$ yd on each side.

    a. Find the total area of the basil plot.

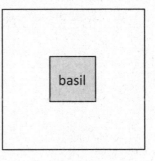

    b. Mrs. Johnson puts a fence around the basil. If the fence is 2 ft from the edge of the garden on each side, what is the perimeter of the fence in feet?

Lesson 15:   Solve real-world problems involving area of figures with fractional side lengths using visual models and/or equations.

©2015 Great Minds. eureka-math.org
G5-M5-SE-BK3-1.3.1-02.2016

85

c.   What is the total area, in square feet, that the fence encloses?

3.   Janet bought 5 yards of fabric $2\frac{1}{4}$-feet wide to make curtains. She used $\frac{1}{3}$ of the fabric to make a long set of curtains and the rest to make 4 short sets.

a.   Find the area of the fabric she used for the long set of curtains.

b.   Find the area of the fabric she used for each of the short sets.

Lesson 15:   Solve real-world problems involving area of figures with fractional side lengths using visual models and/or equations.

©2015 Great Minds. eureka-math.org
G5-M5-SE-BK3-1.3.1-02.2016

4. Some wire is used to make 3 rectangles: A, B, and C. Rectangle B's dimensions are $\frac{3}{5}$ cm larger than Rectangle A's dimensions, and Rectangle C's dimensions are $\frac{3}{5}$ cm larger than Rectangle B's dimensions. Rectangle A is 2 cm by $3\frac{1}{5}$ cm.

a. What is the total area of all three rectangles?

b. If a 40-cm coil of wire was used to form the rectangles, how much wire is left?

Lesson 15: Solve real-world problems involving area of figures with fractional side lengths using visual models and/or equations.

©2015 Great Minds. eureka-math.org
G5-M5-SE-BK3-1.3.1-02.2016

87

This page intentionally left blank

Name _____ Date _____

1. The width of a picnic table is 3 times its length. If the length is $\frac{5}{6}$-yd long, what is the area of the picnic table in square feet?

2. A painting company will paint this wall of a building. The owner gives them the following dimensions:

Window A is $6\frac{1}{4}$ ft × $5\frac{3}{4}$ ft.

Window B is $3\frac{1}{8}$ ft × 4 ft.

Window C is $9\frac{1}{2}$ ft².

Door D is 4 ft × 8 ft.

What is the area of the painted part of the wall?

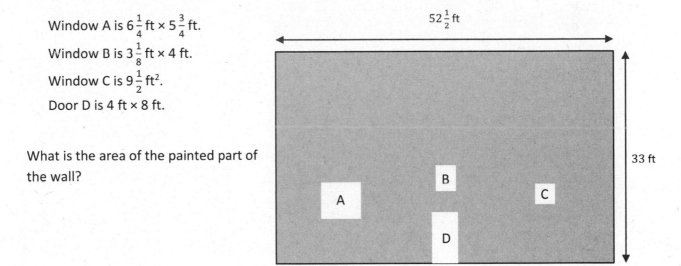

3. The width of a picnic table is 3 times its length.

Lesson 15: Solve real-world problems involving area of figures with fractional side lengths using visual models and/or equations.

©2015 Great Minds. eureka-math.org
G5-M5-SE-BK3-1.3.1-02.2016

89

3. A decorative wooden piece is made up of four rectangles as shown to the right. The smallest rectangle measures $4\frac{1}{2}$ inches by $7\frac{3}{4}$ inches. If $2\frac{1}{4}$ inches are added to each dimension as the rectangles get larger, what is the total area of the entire piece?

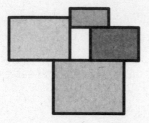

Lesson 15: Solve real-world problems involving area of figures with fractional side lengths using visual models and/or equations.

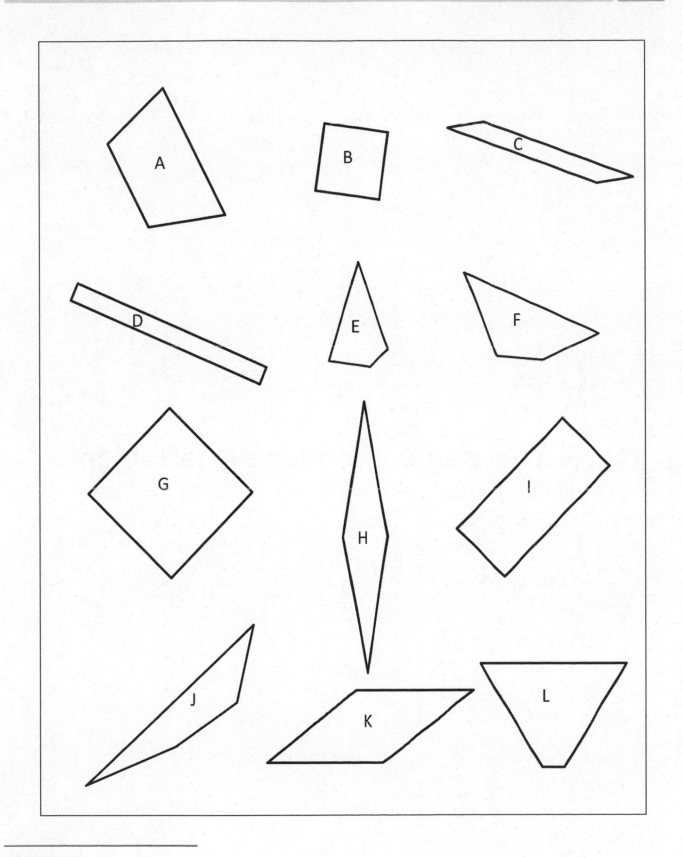

shape sheet

Lesson 15:   Solve real-world problems involving area of figures with fractional side
lengths using visual models and/or equations.

91

©2015 Great Minds. eureka-math.org
G5-M5-SE-BK3-1.3.1-02.2016

This page intentionally left blank

Name _____   Date _____

1. Draw a pair of parallel lines in each box. Then, use the parallel lines to draw a trapezoid with the following:

a. No right angles.	b. Only 1 obtuse angle.
c. 2 obtuse angles.	d. At least 1 right angle.

2. Use the trapezoids you drew to complete the tasks below.

   a. Measure the angles of the trapezoid with your protractor, and record the measurements on the figures.

   b. Use a marker or crayon to circle pairs of angles inside each trapezoid with a sum equal to 180°. Use a different color for each pair.

3. List the properties that are shared by all the trapezoids that you worked with today.

4. When can a quadrilateral also be called a trapezoid?

5. Follow the directions to draw one last trapezoid.

   a. Draw a segment $\overline{AB}$ parallel to the bottom of this page that is 5 cm long.

   b. Draw two 55° angles with vertices at A and B so that an isosceles triangle is formed with $\overline{AB}$ as the base of the triangle.

   c. Label the top vertex of your triangle as C.

   d. Use your set square to draw a line parallel to $\overline{AB}$ that intersects both $\overline{AC}$ and $\overline{BC}$.

   e. Shade the trapezoid that you drew.

Lesson 16:   Draw trapezoids to clarify their attributes, and define trapezoids based on those attributes.

©2015 Great Minds. eureka-math.org
G5-M5-SE-BK3-1.3.1-02.2016

Name _____   Date _____

1. Use a straightedge and the grid paper to draw:

    a. A trapezoid with exactly 2 right angles.   b. A trapezoid with no right angles.

2. Kaplan incorrectly sorted some quadrilaterals into trapezoids and non-trapezoids as pictured below.

    a. Circle the shapes that are in the wrong group, and tell why they are sorted incorrectly.

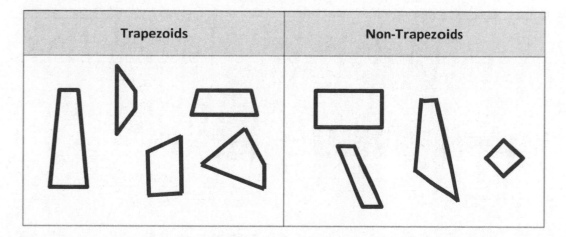

Trapezoids	Non-Trapezoids

    b. Explain what tools would be necessary to use to verify the placement of all the trapezoids.

Lesson 16:   Draw trapezoids to clarify their attributes, and define trapezoids based   95
                   on those attributes.

©2015 Great Minds. eureka-math.org
G5-M5-SE-BK3-1.3.1-02.2016

3.   a.   Use a straightedge to draw an isosceles trapezoid on the grid paper.

   b.   Why is this shape called an isosceles trapezoid?

Lesson 16:   Draw trapezoids to clarify their attributes, and define trapezoids based on those attributes.

©2015 Great Minds. eureka-math.org
G5-M5-SE-BK3-1.3.1-02.2016

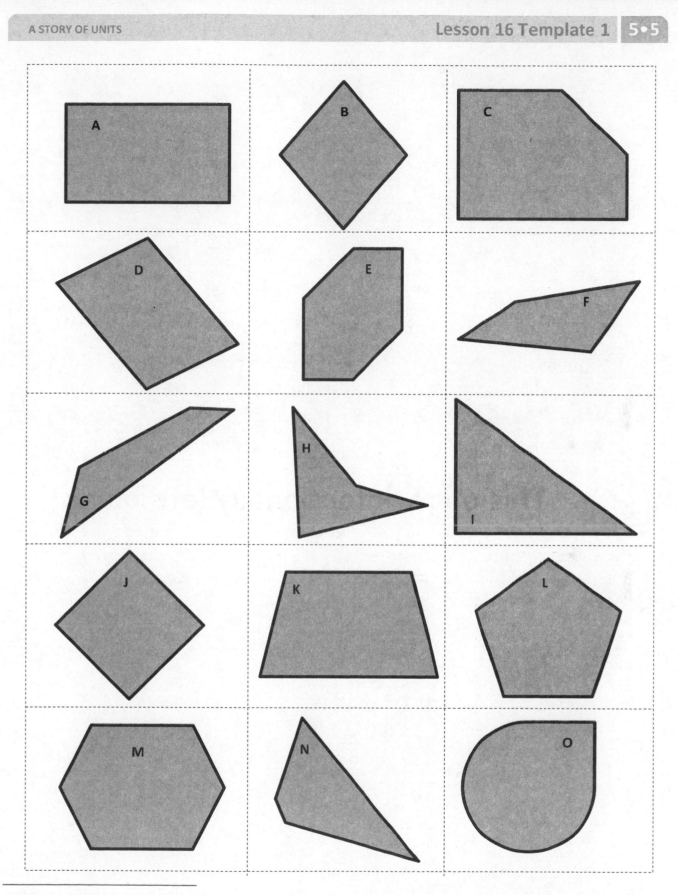

collection of polygons

Lesson 16:   Draw trapezoids to clarify their attributes, and define trapezoids based
on those attributes.

97

©2015 Great Minds. eureka-math.org
G5-M5-SE-BK3-1.3.1-02.2016

This page intentionally left blank

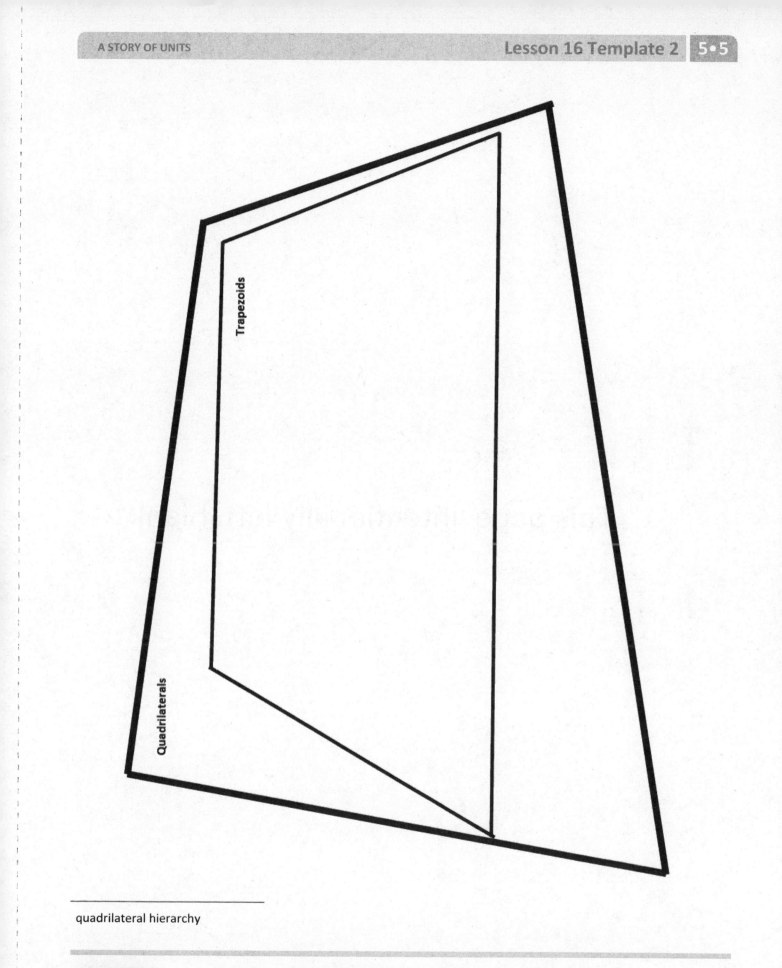

Trapezoids

Quadrilaterals

quadrilateral hierarchy

EUREKA
MATH™

Lesson 16: Draw trapezoids to clarify their attributes, and define trapezoids based
on those attributes.

©2015 Great Minds. eureka-math.org
G5-M5-SE-BK3-1.3.1-02.2016

99

This page intentionally left blank

Name _____ Date _____

1. Draw a parallelogram in each box with the attributes listed.

a. No right angles.	b. At least 2 right angles.
c. Equal sides with no right angles.	d. All sides equal with at least 2 right angles.

Lesson 17:   Draw parallelograms to clarify their attributes, and define parallelograms based on those attributes.

101

©2015 Great Minds. eureka-math.org
G5-M5-SE-BK3-1.3.1-02.2016

2. Use the parallelograms you drew to complete the tasks below.

   a. Measure the angles of the parallelogram with your protractor, and record the measurements on the figures.

   b. Use a marker or crayon to circle pairs of angles inside each parallelogram with a sum equal to 180°. Use a different color for each pair.

3. Draw another parallelogram below.

   a. Draw the diagonals, and measure their lengths. Record the measurements to the side of your figure.

   b. Measure the length of each of the four segments of the diagonals from the vertices to the point of intersection of the diagonals. Color the segments that have the same length the same color. What do you notice?

4. List the properties that are shared by all of the parallelograms that you worked with today.

   a. When can a quadrilateral also be called a parallelogram?

   b. When can a trapezoid also be called a parallelogram?

Lesson 17:   Draw parallelograms to clarify their attributes, and define parallelograms based on those attributes.

©2015 Great Minds. eureka-math.org
G5-M5-SE-BK3-1.3.1-02.2016

Name _____   Date _____

1. ∠A measures 60°.

    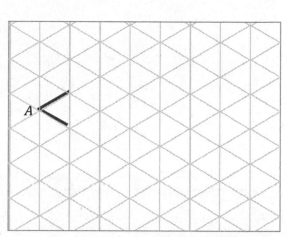

    a. Extend the rays of ∠A, and draw parallelogram ABCD
        on the grid paper.

    b. What are the measures of ∠B, ∠C, and ∠D?

2. WXYZ is a parallelogram not drawn to scale.

    a. Using what you know about parallelograms, give the
        measure of sides XY and YZ.

        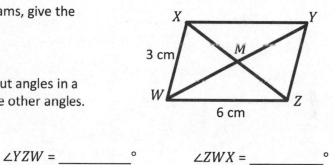

    b. ∠WXY = 113°. Use what you know about angles in a
        parallelogram to find the measure of the other angles.

    ∠XYZ = _____°   ∠YZW = _____°   ∠ZWX = _____°

3. Jack measured some segments in Problem 2. He found that $\overline{WY}$ = 8 cm and $\overline{MZ}$ = 3 cm.

    Give the lengths of the following segments:

    WM = _____ cm   MY = _____ cm

    XM = _____ cm   XZ = _____ cm

Lesson 17:   Draw parallelograms to clarify their attributes, and define   103
               parallelograms based on those attributes.

©2015 Great Minds. eureka-math.org
G5-M5-SE-BK3-1.3.1-02.2016

4. Using the properties of shapes, explain why all parallelograms are trapezoids.

5. Teresa says that because the diagonals of a parallelogram bisect each other, if one diagonal is 4.2 cm, the other diagonal must be half that length. Use words and pictures to explain Teresa's error.

Lesson 17: Draw parallelograms to clarify their attributes, and define parallelograms based on those attributes.

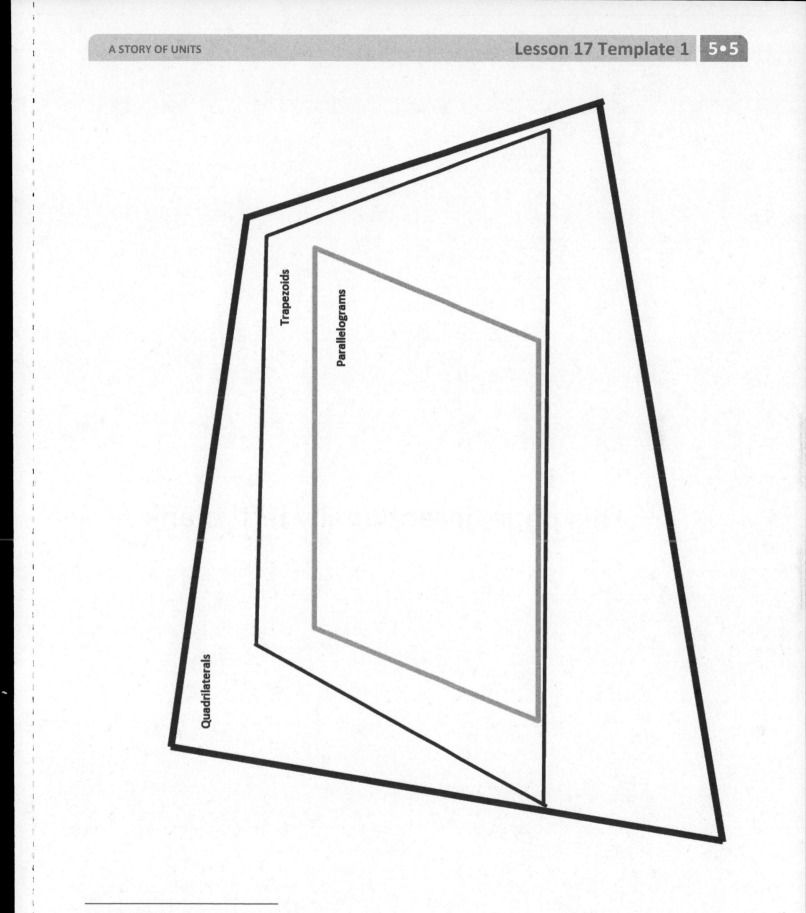

Quadrilaterals

Trapezoids

Parallelograms

quadrilateral hierarchy with parallelogram

Lesson 17:   Draw parallelograms to clarify their attributes, and define
              parallelograms based on those attributes.

105

©2015 Great Minds. eureka-math.org
G5-M5-SE-BK3-1.3.1-02.2016

This page intentionally left blank

Name _____ Date _____

1. Draw the figures in each box with the attributes listed.

a. Rhombus with no right angles	b. Rectangle with not all sides equal
c. Rhombus with 1 right angle	d. Rectangle with all sides equal

2. Use the figures you drew to complete the tasks below.

   a. Measure the angles of the figures with your protractor, and record the measurements on the figures.

   b. Use a marker or crayon to circle pairs of angles inside each figure with a sum equal to 180°. Use a different color for each pair.

Lesson 18:   Draw rectangles and rhombuses to clarify their attributes, and define
              rectangles and rhombuses based on those attributes.

©2015 Great Minds. eureka-math.org
G5-M5-SE-BK3-1.3.1-02.2016

107

3. Draw a rhombus and a rectangle below.

    a. Draw the diagonals, and measure their lengths. Record the measurements on the figure.

    b. Measure the length of each segment of the diagonals from the vertex to the intersection point of the diagonals. Using a marker or crayon, color segments that have the same length. Use a different color for each different length.

4. a. List the properties that are shared by all of the rhombuses that you worked with today.

    b. List the properties that are shared by all of the rectangles that you worked with today.

    c. When can a trapezoid also be called a rhombus?

    d. When can a parallelogram also be called a rectangle?

    e. When can a quadrilateral also be called a rhombus?

Lesson 18:   Draw rectangles and rhombuses to clarify their attributes, and define rectangles and rhombuses based on those attributes.

©2015 Great Minds. eureka-math.org
G5-M5-SE-BK3-1.3.1-02.2016

Name _____   Date _____

1. Use the grid paper to draw.

   a. A rhombus with no right angles

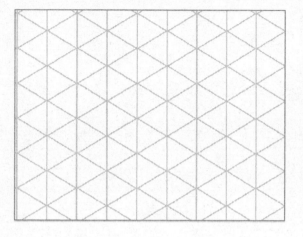

   b. A rhombus with 4 right angles

   c. A rectangle with not all sides equal

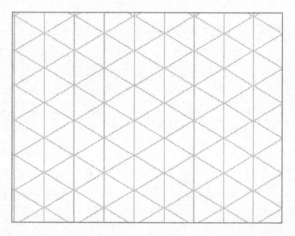

   d. A rectangle with all sides equal

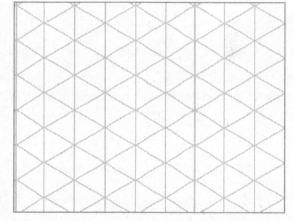

Lesson 18:   Draw rectangles and rhombuses to clarify their attributes, and define
rectangles and rhombuses based on those attributes.

109

©2015 Great Minds. eureka-math.org
G5-M5-SE-BK3-1.3.1-02.2016

2. A rhombus has a perimeter of 217 cm. What is the length of each side of the rhombus?

3. List the properties that all rhombuses share.

4. List the properties that all rectangles share.

Lesson 18: Draw rectangles and rhombuses to clarify their attributes, and define rectangles and rhombuses based on those attributes.

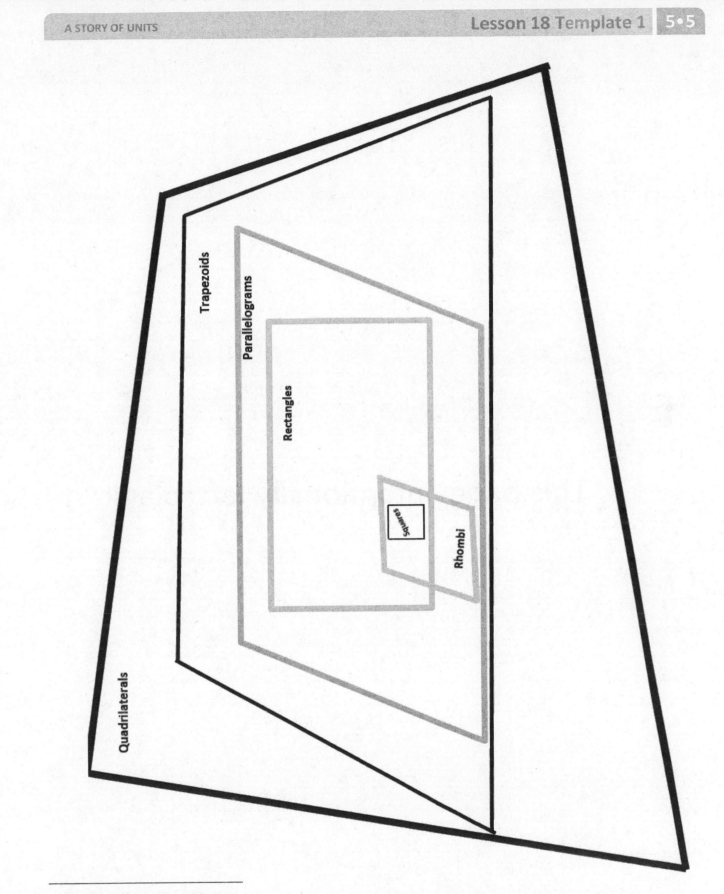

quadrilateral hierarchy with square

Lesson 18:   Draw rectangles and rhombuses to clarify their attributes, and define rectangles and rhombuses based on those attributes.

111

©2015 Great Minds. eureka-math.org
G5-M5-SE-BK3-1.3.1-02.2016

This page intentionally left blank

Name _____   Date _____

1. Draw the figures in each box with the attributes listed. If your figure has more than one name, write it in the box.

a. Rhombus with 2 right angles	b. Kite with all sides equal
c. Kite with 4 right angles	d. Kite with 2 pairs of adjacent sides equal (The pairs are not equal to each other.)

2. Use the figures you drew to complete the tasks below.

   a. Measure the angles of the figures with your protractor, and record the measurements on the figures.

   b. Use a marker or crayon to circle pairs of angles that are equal in measure, inside each figure. Use a different color for each pair.

EUREKA MATH

Lesson 19:   Draw kites and squares to clarify their attributes, and define kites and squares based on those attributes.

©2015 Great Minds. eureka-math.org
G5-M5-SE-BK3-1.3.1-02.2016

113

3. a. List the properties shared by all of the squares that you worked with today.

b. List the properties shared by all of the kites that you worked with today.

c. When can a rhombus also be called a square?

d. When can a kite also be called a square?

e. When can a trapezoid also be called a kite?

Lesson 19:   Draw kites and squares to clarify their attributes, and define kites and squares based on those attributes.

©2015 Great Minds. eureka-math.org
G5-M5-SE-BK3-1.3.1-02.2016

Name _____   Date _____

1. a.   Draw a kite that is not a parallelogram on the grid paper.

    b.   List all the properties of a kite.

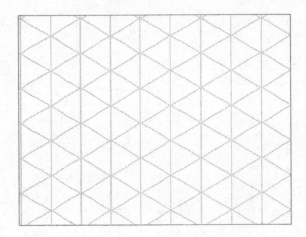

    c.   When can a parallelogram also be a kite?

2. If rectangles must have right angles, explain how a rhombus could also be called a rectangle.

3. Draw a rhombus that is also a rectangle on the grid paper.

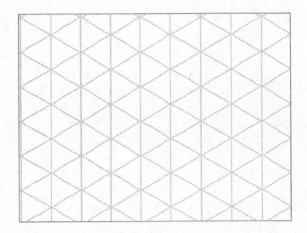

Lesson 19:   Draw kites and squares to clarify their attributes, and define kites and
             squares based on those attributes.

©2015 Great Minds. eureka-math.org
G5-M5-SE-BK3-1.3.1-02.2016

115

4. Kirkland says that figure *EFGH* below is a quadrilateral because it has four points in the same plane and four segments with no three endpoints collinear. Explain his error.

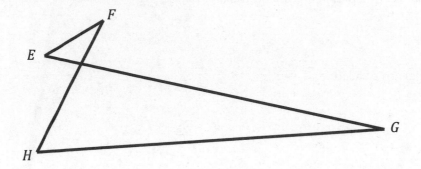

©2015 Great Minds. eureka-math.org
G5-M5-SE-BK3-1.3.1-02.2016

EUREKA
MATH

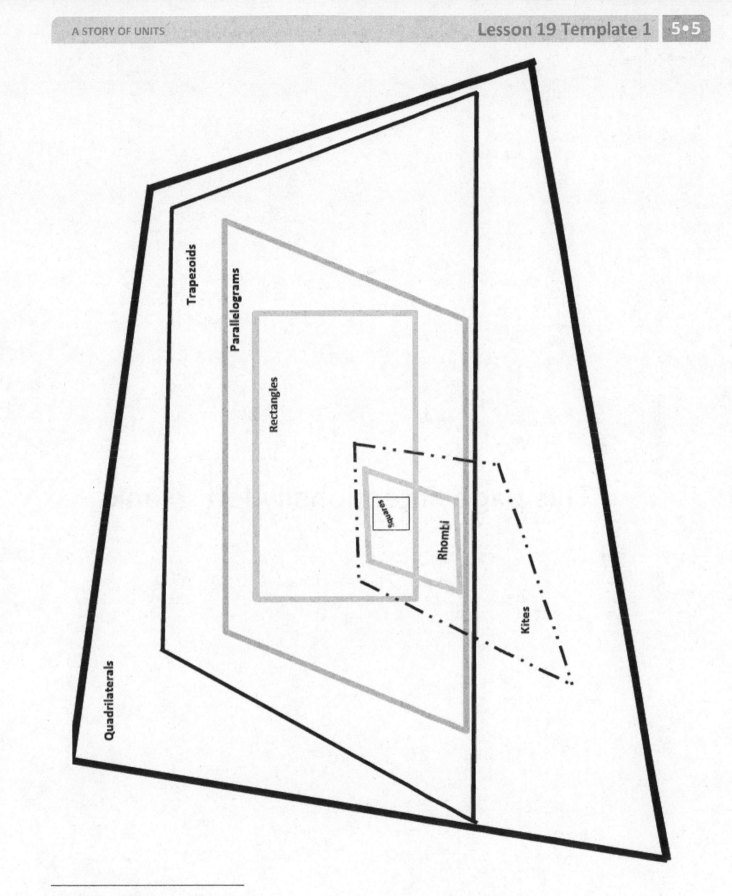

quadrilateral hierarchy with kite

Lesson 19: Draw kites and squares to clarify their attributes, and define kites and squares based on those attributes.

117

©2015 Great Minds. eureka-math.org
G5-M5-SE-BK3-1.3.1-02.2016

This page intentionally left blank

Name _____ Date _____

1. True or false. If the statement is false, rewrite it to make it true.

		T	F
a.	All trapezoids are quadrilaterals.		
b.	All parallelograms are rhombuses.		
c.	All squares are trapezoids.		
d.	All rectangles are squares.		
e.	Rectangles are always parallelograms.		
f.	All parallelograms are trapezoids.		
g.	All rhombuses are rectangles.		
h.	Kites are never rhombuses.		
i.	All squares are kites.		
j.	All kites are squares.		
k.	All rhombuses are squares.		

2. Fill in the blanks.

a. *ABCD* is a trapezoid. Find the measurements listed below.

∠A = _____°

∠D = _____°

What other names does this figure have?

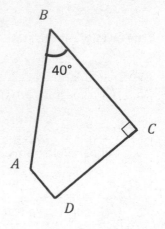

b. *RECT* is a rectangle. Find the measurements listed below.

Line *TE* = _____

Line *RC* = _____

Line *CT* = _____

∠ERM = _____°

∠CTR = _____°

What other names does this figure have?

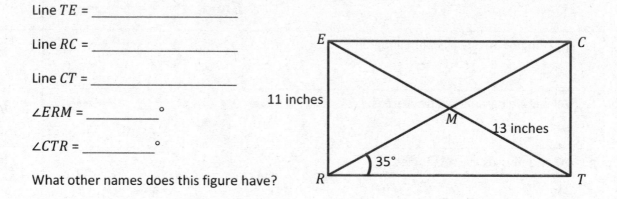

c. *PARL* is a parallelogram. Find the measurements listed below.

Line *AL* = _____

Line *PR* = _____

∠ARL = _____°

∠PAR = _____°

∠RLP = _____°

What other names does this figure have?

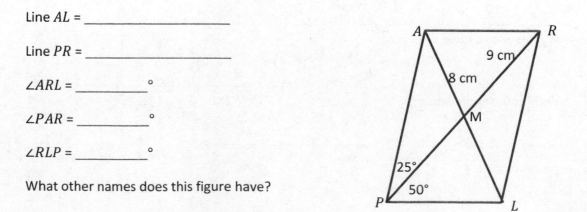

Lesson 20: Classify two-dimensional figures in a hierarchy based on properties.

©2015 Great Minds. eureka-math.org
G5-M5-SE-BK3-1.3.1-02.2016

EUREKA
MATH

Name _____ Date _____

1. Follow the flow chart, and put the name of the figure in the boxes.

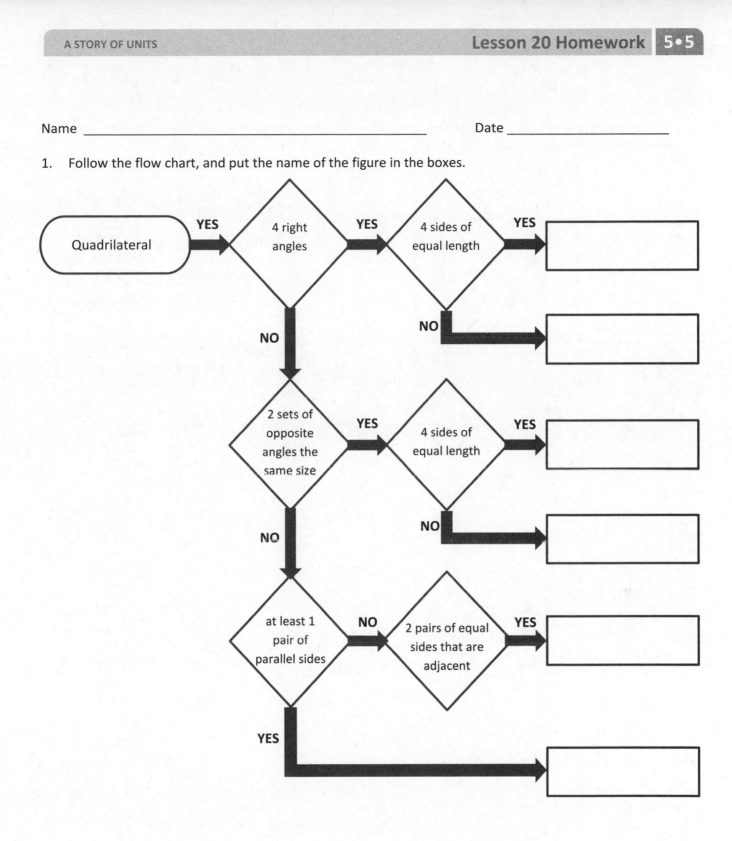

EUREKA
MATH

Lesson 20:   Classify two-dimensional figures in a hierarchy based on properties.

121

©2015 Great Minds. eureka-math.org
G5-M5-SE-BK3-1.3.1-02.2016

2. *SQRE* is a square with an area of 49 cm², and *RM* = 4.95 cm. Find the measurements using what you know about the properties of squares.

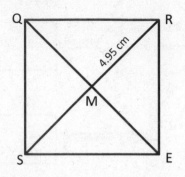

a.   *RS* = _____ cm

b.   *QE* = _____ cm

c.   Perimeter = _____ cm

d. $m\angle QRE$ = _____ °

e. $m\angle RMQ$ = _____ °

©2015 Great Minds. eureka-math.org
G5-M5-SE-BK3-1.3.1-02.2016

Quadrilaterals	**Trapezoids**
Parallelograms	**Rectangles**
Rhombuses	**Kites**
Squares	**Polygons**

shape name cards

This page intentionally left blank

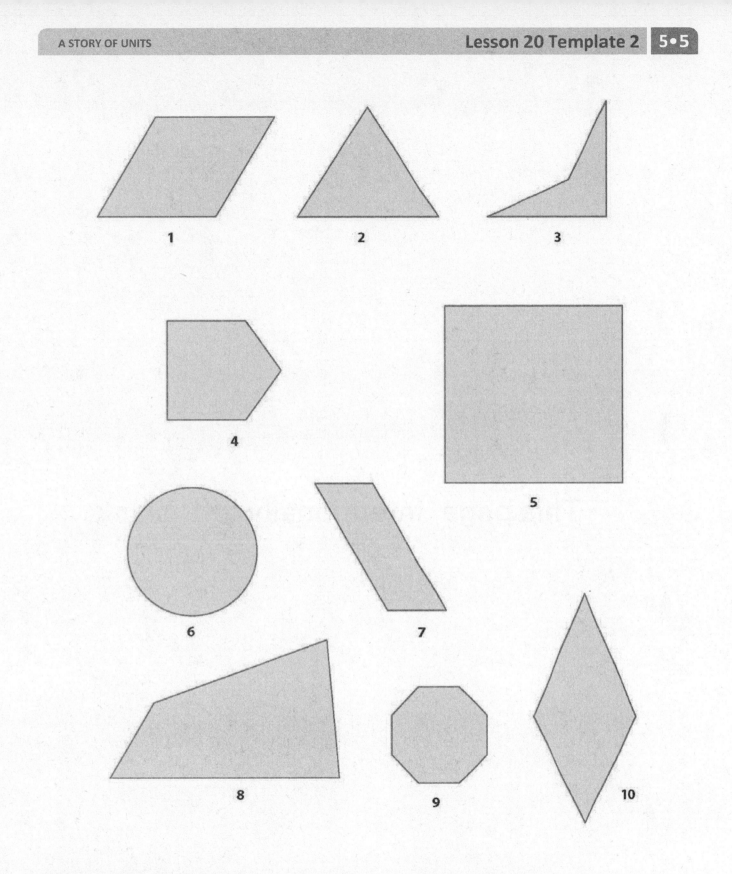

shapes for sorting (page 1)

Lesson 20:   Classify two-dimensional figures in a hierarchy based on properties.

125

©2015 Great Minds. eureka-math.org
G5-M5-SE-BK3-1.3.1-02.2016

This page intentionally left blank

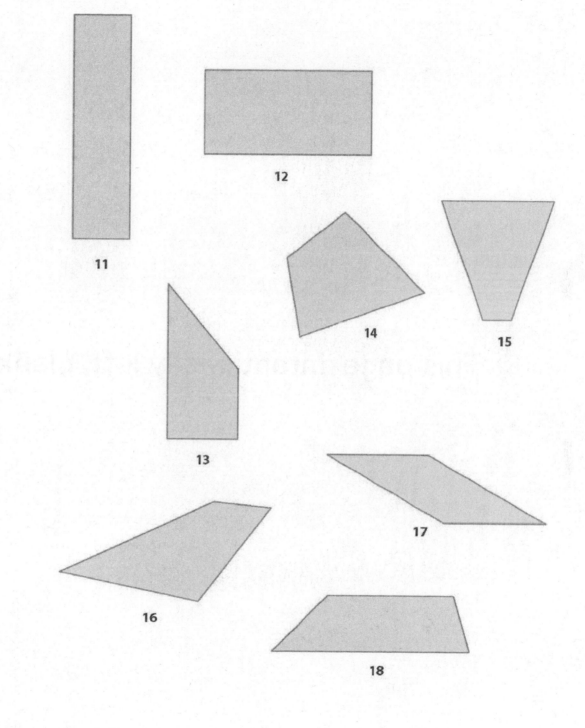

11

12

13

14

15

16

17

18

shapes for sorting (page 2)

Lesson 20:   Classify two-dimensional figures in a hierarchy based on properties.

127

©2015 Great Minds. eureka-math.org
G5-M5-SE-BK3-1.3.1-02.2016

This page intentionally left blank

Name _____ Date _____

1. Write the number on your task card and a summary of the task in the blank. Then, draw the figure in the box. Label your figure with as many names as you can. Circle the most specific name.

Task # ___: _____

Task # ___: _____

Task # ___: _____

Task # ___: _____

Task # ___: _____

Task # ___: _____

Lesson 21: Draw and identify varied two-dimensional figures from given attributes.

129

©2015 Great Minds. eureka-math.org
G5-M5-SE-BK3-1.3.1-02.2016

2. John says that because rhombuses do not have perpendicular sides, they cannot be rectangles. Explain his error in thinking.

3. Jack says that because kites do not have parallel sides, a square is not a kite. Explain his error in thinking.

Lesson 21: Draw and identify varied two-dimensional figures from given attributes.

©2015 Great Minds. eureka-math.org
G5-M5-SE-BK3-1.3.1-02.2016

Name _____ Date _____

1. Answer the questions by checking the box.

	Sometimes	Always
a. Is a square a rectangle?		
b. Is a rectangle a kite?		
c. Is a rectangle a parallelogram?		
d. Is a square a trapezoid?		
e. Is a parallelogram a trapezoid?		
f. Is a trapezoid a parallelogram?		
g. Is a kite a parallelogram?		

   h. For each statement that you answered with *sometimes*, draw and label an example that justifies your answer.

2. Use what you know about quadrilaterals to answer each question below.

   a. Explain when a trapezoid is not a parallelogram. Sketch an example.

   b. Explain when a kite is not a parallelogram. Sketch an example.

Lesson 21: Draw and identify varied two-dimensional figures from given attributes.

©2015 Great Minds. eureka-math.org
G5-M5-SE-BK3-1.3.1-02.2016

131

This page intentionally left blank

Task 3:
Draw a quadrilateral with 2 pairs of equal sides and no parallel sides.

Task 6:
Draw a rhombus with 4 equal angles.

Task 2:
Draw a rectangle with a length that is twice its width.

Task 5:
Draw a parallelogram with two pairs of perpendicular sides.

Task 1:
Draw a trapezoid with a right angle.

Task 4:
Draw a rhombus with right angles.

task cards (1–6)

Lesson 21:   Draw and identify varied two-dimensional figures from given attributes.

©2015 Great Minds. eureka-math.org
G5-M5-SE-BK3-1.3.1-02.2016

133

This page intentionally left blank

Task 7: Draw a quadrilateral with four equal sides.	Task 8: Draw a parallelogram with right angles.	Task 9: Draw a parallelogram with a side of 4 cm and a side of 6 cm.
Task 10: Draw an isosceles trapezoid.	Task 11: Draw a parallelogram with no right angles.	Task 12: Draw a rectangle that is also a rhombus.

task cards (7–12)

Lesson 21:   Draw and identify varied two-dimensional figures from given attributes.

135

This page intentionally left blank

Task 15:
Draw a trapezoid with four right angles.

Task 18:
Draw a rectangle that is not a rhombus.

Task 14:
Draw a quadrilateral that has only one pair of equal opposite angles.

Task 17:
Draw a parallelogram with a 60° angle.

Task 13:
Draw a quadrilateral that has at least one pair of equal opposite angles.

Task 16:
Draw a kite that is also a parallelogram.

task cards (13–18)

Lesson 21:   Draw and identify varied two-dimensional figures from given attributes.

©2015 Great Minds. eureka-math.org
G5-M5-SE-BK3-1.3.1-02.2016

This page intentionally left blank

Task 21:
Draw a kite that is not a parallelogram.

Task 24:
Draw a quadrilateral whose diagonals do not bisect each other.

Task 20:
Draw a parallelogram that is not a rectangle.

Task 23:
Draw a trapezoid that is not a parallelogram.

Task 19:
Draw a rhombus that is not a rectangle.

Task 22:
Draw a quadrilateral whose diagonals bisect each other at a right angle.

task cards (19–24)

This page intentionally left blank

Eureka Math
Grade 5
Module 6

Special thanks go to the Gordon A. Cain Center and to the Department of Mathematics at Louisiana State University for their support in the development of *Eureka Math*.

For a free *Eureka Math* Teacher
Resource Pack, Parent Tip
Sheets, and more please
visit www.Eureka.tools

Published by the non-profit Great Minds

Copyright © 2015 Great Minds. No part of this work may be reproduced, sold, or commercialized, in whole or in part, without written permission from Great Minds. Non-commercial use is licensed pursuant to a Creative Commons Attribution-NonCommercial-ShareAlike 4.0 license; for more information, go to http://greatminds.net/maps/math/copyright. "Great Minds" and "Eureka Math" are registered trademarks of Great Minds.

Printed in the U.S.A.
This book may be purchased from the publisher at eureka-math.org
10 9 8

ISBN 978-1-63255-310-2

Name _____ Date _____

1. Each shape was placed at a point on the number line s. Give the coordinate of each point below.

   a. 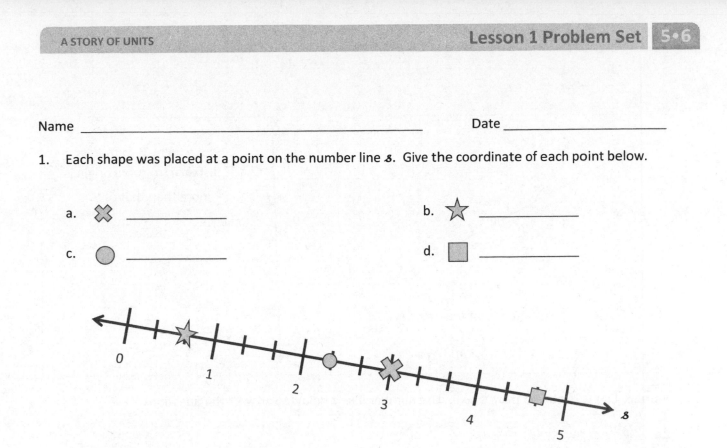 _____

   b. ☆ _____

   c. ⬤ _____

   d. ▢ _____

2. Plot the points on the number lines.

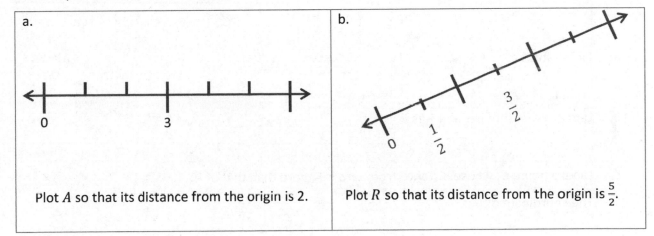

   a.

   Plot A so that its distance from the origin is 2.

   b.

   Plot R so that its distance from the origin is $\frac{5}{2}$.

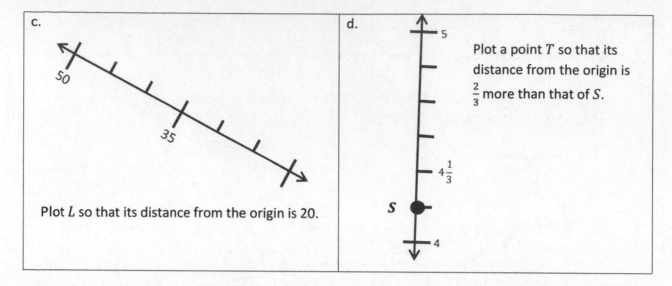

c.

50

35

Plot L so that its distance from the origin is 20.

d.

Plot a point T so that its distance from the origin is $\frac{2}{3}$ more than that of S.

5

$4\frac{1}{3}$

S

4

3. Number line g is labeled from 0 to 6. Use number line g below to answer the questions.

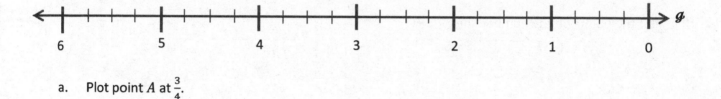

6   5   4   3   2   1   0 g

a. Plot point A at $\frac{3}{4}$.

b. Label a point that lies at $4\frac{1}{2}$ as B.

c. Label a point, C, whose distance from zero is 5 more than that of A.

   The coordinate of C is _____.

d. Plot a point, D, whose distance from zero is $1\frac{1}{4}$ less than that of B.

   The coordinate of D is _____.

e. The distance of E from zero is $1\frac{3}{4}$ more than that of D. Plot point E.

f. What is the coordinate of the point that lies halfway between A and D? _____
   Label this point F.

EUREKA
MATH

4. Mrs. Fan asked her fifth-grade class to create a number line. Lenox created the number line below:

    12   10   8   6   4   2   0

    Parks said Lenox's number line is wrong because numbers should always increase from left to right. Who is correct? Explain your thinking.

5. A pirate marked the palm tree on his treasure map and buried his treasure 30 feet away. Do you think he will be able to easily find his treasure when he returns? Why or why not? What might he do to make it easier to find?

Look for the treasure 30 feet from this tree!

This page intentionally left blank

Name _____   Date _____

1. Answer the following questions using number line q below.

    a. What is the coordinate, or the distance from the origin, of the 😊 ? _____

    b. What is the coordinate of the ⚡ ? _____

    c. What is the coordinate of the 🤍 ? _____

    d. What is the coordinate at the midpoint of the ⚡ and the 🤍 ? _____

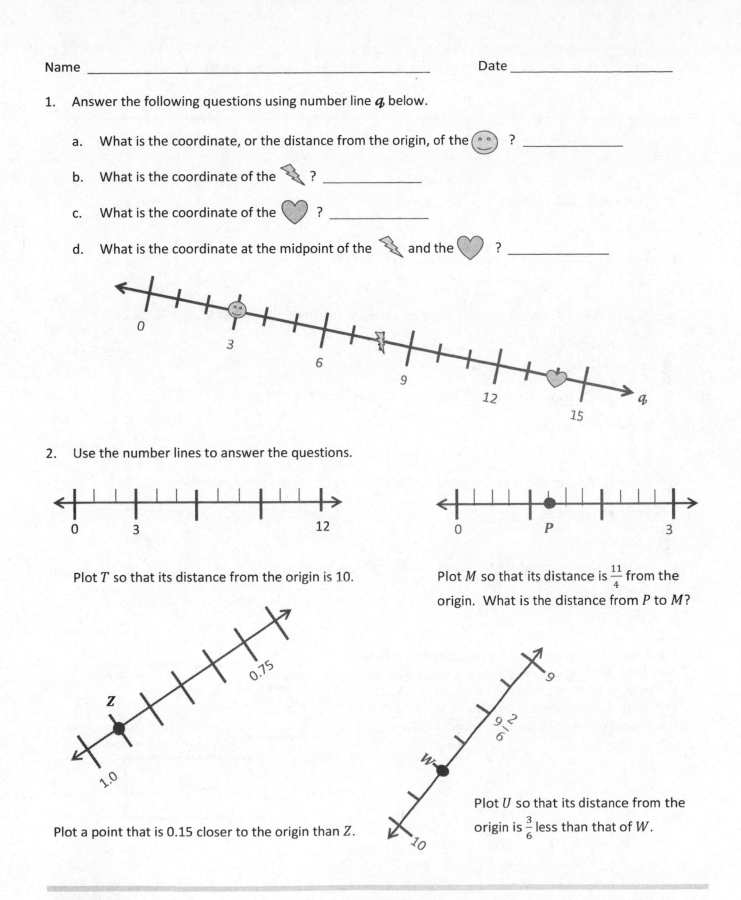

2. Use the number lines to answer the questions.

Plot T so that its distance from the origin is 10.

Plot M so that its distance is $\frac{11}{4}$ from the origin. What is the distance from P to M?

Plot a point that is 0.15 closer to the origin than Z.

Plot U so that its distance from the origin is $\frac{3}{6}$ less than that of W.

3. Number line k shows 12 units. Use number line k below to answer the questions.

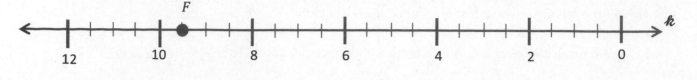

a. Plot a point at 1. Label it A.

b. Label a point that lies at $3\frac{1}{2}$ as B.

c. Label a point, C, whose distance from zero is 8 units farther than that of B.

    The coordinate of C is _____.

d. Plot a point, D, whose distance from zero is $\frac{6}{2}$ less than that of B.

    The coordinate of D is _____.

e. What is the coordinate of the point that lies $\frac{17}{2}$ farther from the origin than D?

    Label this point E.

f. What is the coordinate of the point that lies halfway between F and D?

    Label this point G.

4. Mr. Baker's fifth-grade class buried a time capsule in the
    field behind the school. They drew a map and marked
    the location of the capsule with an ✖ so that his class
    can dig it up in ten years. What could Mr. Baker's class
    have done to make the capsule easier to find?

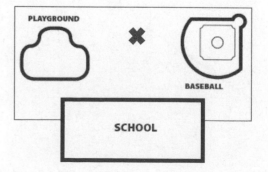

Lesson 1:   Construct a coordinate system on a line.

EUREKA
MATH™

Name _____   Date _____

1.

a. Use a set square to draw a line perpendicular to the x-axes through points P, Q, and R. Label the new line as the y-axis.

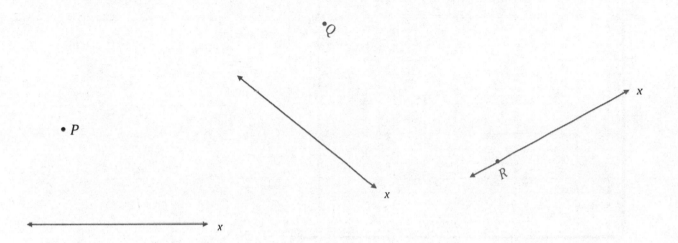

a. Choose one of the sets of perpendicular lines above, and create a coordinate plane. Mark 7 units on each axis, and label them as whole numbers.

2. Use the coordinate plane to answer the following.

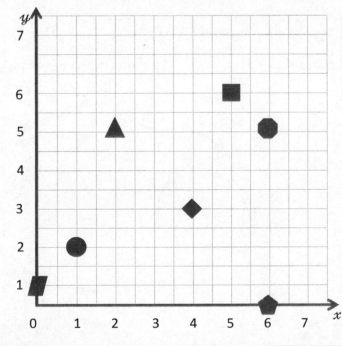

a. Name the shape at each location.

x-coordinate	y-coordinate	Shape
2	5	
1	2	
5	6	
6	5	

b. Which shape is 2 units from the y-axis?

c. Which shape has an x-coordinate of 0?

d. Which shape is 4 units from the y-axis and 3 units from the x-axis?

3. Use the coordinate plane to answer the following.

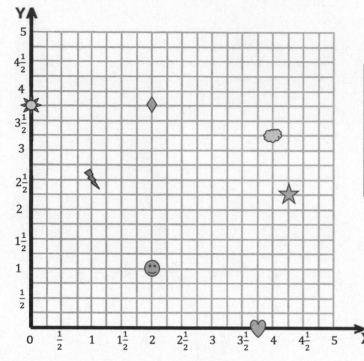

a. Fill in the blanks.

Shape	x-coordinate	y-coordinate
Smiley Face		
Diamond		
Sun		
Heart		

b. Name the shape whose x-coordinate is $\frac{1}{2}$ more than the value of the heart's x-coordinate.

c. Plot a triangle at (3, 4).   d. Plot a square at $(4\frac{3}{4}, 5)$.   e. Plot an X at $(\frac{1}{2}, \frac{3}{4})$.

4. The pirate's treasure is buried at the ✖ on the map. How could a coordinate plane make describing its location easier?

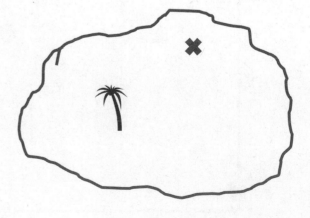

Lesson 2:   Construct a coordinate system on a plane.   **EUREKA MATH**

Name _____   Date _____

1.

a.   Use a set square to draw a line perpendicular to the x-axis through point P. Label the new line as the y-axis.

b.   Choose one of the sets of perpendicular lines above, and create a coordinate plane. Mark 5 units on each axis, and label them as whole numbers.

2.   Use the coordinate plane to answer the following.

a.   Name the shape at each location.

x-coordinate	y-coordinate	Shape
2	4	
5	4	
1	5	
5	1	

b.   Which shape is 2 units from the x-axis?

c.   Which shape has the same x- and y-coordinate?

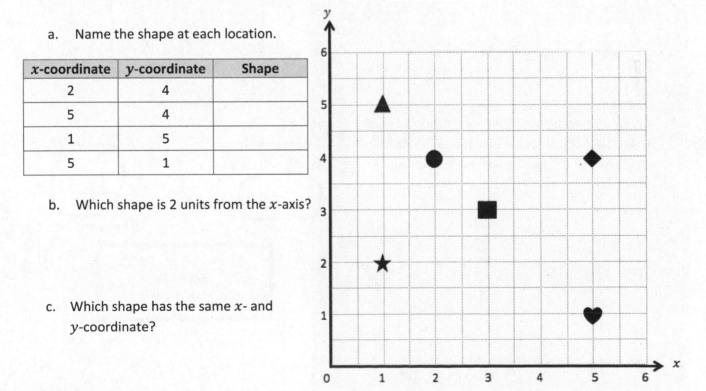

3. Use the coordinate plane to answer the following.

   a. Name the coordinates of each shape.

Shape	x-coordinate	y-coordinate
Moon		
Sun		
Heart		
Cloud		
Smiley Face		

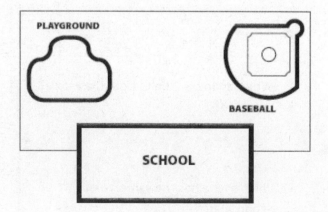

   b. Which 2 shapes have the same y-coordinate?

   c. Plot an X at (2, 3).

   d. Plot a square at $(3, 2\frac{1}{2})$.

   e. Plot a triangle at $(6, 3\frac{1}{2})$.

4. Mr. Palmer plans to bury a time capsule 10 yards behind the school. What else should he do to make naming the location of the time capsule more accurate?

Lesson 2:   Construct a coordinate system on a plane.

©2015 Great Minds eureka-math.org
G5-M6-SE-BK3-1.3.1-02.2016

EUREKA
MATH™

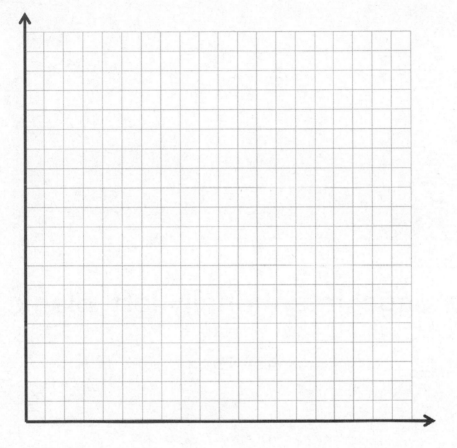

coordinate plane

This page intentionally left blank

Name _____ Date _____

1. Use the grid below to complete the following tasks.

   a. Construct an x-axis that passes through points A and B.

   b. Construct a perpendicular y-axis that passes through points C and F.

   c. Label the origin as 0.

   d. The x-coordinate of B is $5\frac{2}{3}$. Label the whole numbers along the x-axis.

   e. The y-coordinate of C is $5\frac{1}{3}$. Label the whole numbers along the y-axis.

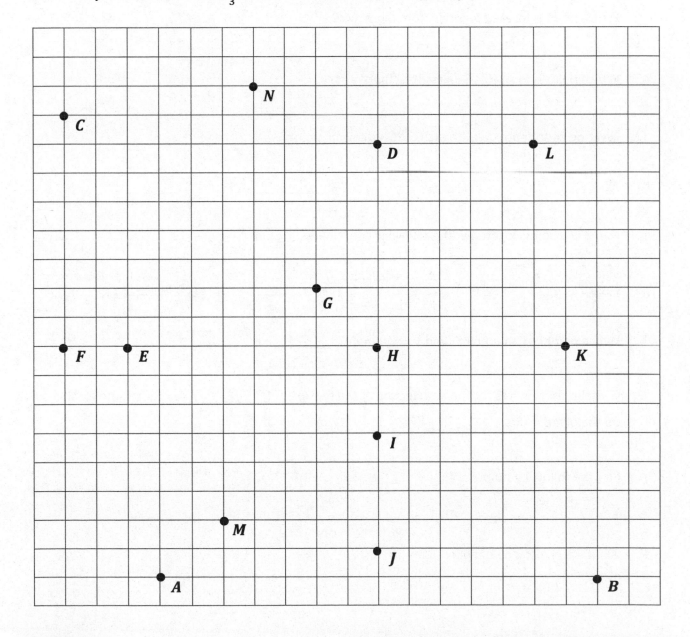

Lesson 3: Name points using coordinate pairs, and use the coordinate pairs to plot points.

©2015 Great Minds eureka-math.org
G5-M6-SE-BK3-1.3.1-02.2016

13

2. For all of the following problems, consider the points A through N on the previous page.

a. Identify all of the points that have an x-coordinate of $3\frac{1}{3}$.

b. Identify all of the points that have a y-coordinate of $2\frac{2}{3}$.

c. Which point is $3\frac{1}{3}$ units above the x-axis *and* $2\frac{2}{3}$ units to the right of the y-axis? Name the point, and give its coordinate pair.

d. Which point is located $5\frac{1}{3}$ units from the y-axis?

e. Which point is located $1\frac{2}{3}$ units along the x-axis?

f. Give the coordinate pair for each of the following points.

K: _____ I: _____ B: _____ C: _____

g. Name the points located at the following coordinates.

$(1\frac{2}{3}, \frac{2}{3})$ _____ $(0, 2\frac{2}{3})$ _____ $(1, 0)$ _____ $(2, 5\frac{2}{3})$ _____

h. Which point has an equal x- and y-coordinate? _____

i. Give the coordinates for the intersection of the two axes. (____ , ____) Another name for this point on the plane is the _____.

j. Plot the following points.

P: $(4\frac{1}{3}, 4)$ Q: $(\frac{1}{3}, 6)$ R: $(4\frac{2}{3}, 1)$ S: $(0, 1\frac{2}{3})$

k. What is the distance between E and H, or EH?

Lesson 3:   Name points using coordinate pairs, and use the coordinate pairs to plot points.

©2015 Great Minds eureka-math.org
G5-M6-SE-BK3-1.3.1-02.2016

l.   What is the length of HD?

m.   Would the length of ED be greater or less than $EH + HD$?

n.   Jack was absent when the teacher explained how to describe the location of a point on the coordinate plane. Explain it to him using point J.

Lesson 3:   Name points using coordinate pairs, and use the coordinate pairs to plot points.

©2015 Great Minds eureka-math.org
G5-M6-SE-BK3-1.3.1-02.2016

15

This page intentionally left blank

Name _____   Date _____

1. Use the grid below to complete the following tasks.

    a. Construct a y-axis that passes through points Y and Z.

    b. Construct a perpendicular x-axis that passes through points Z and X.

    c. Label the origin as 0.

    d. The y-coordinate of W is $2\frac{3}{5}$. Label the whole numbers along the y-axis.

    e. The x-coordinate of V is $2\frac{2}{5}$. Label the whole numbers along the x-axis.

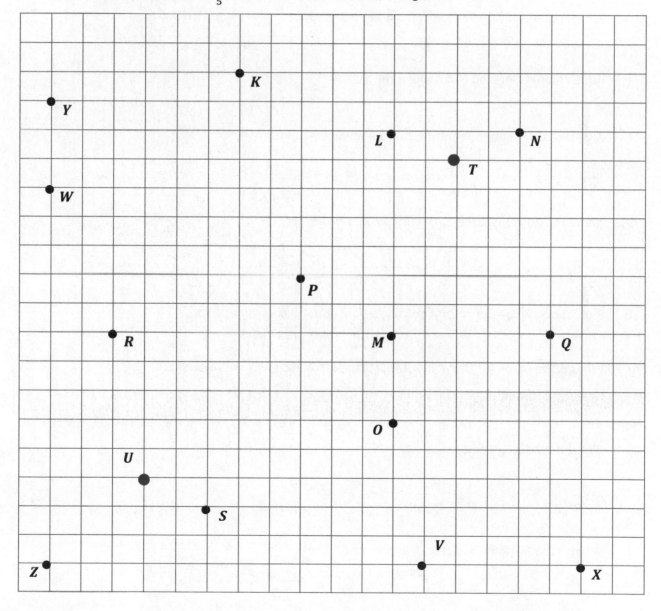

Lesson 3:   Name points using coordinate pairs, and use the coordinate pairs to plot points.

©2015 Great Minds eureka-math.org
G5-M6-SE-BK3-1.3.1-02.2016

17

2. For all of the following problems, consider the points K through X on the previous page.

    a. Identify all of the points that have a y-coordinate of $1\frac{3}{5}$.

    b. Identify all of the points that have an x-coordinate of $2\frac{1}{5}$.

    c. Which point is $1\frac{3}{5}$ units above the x-axis *and* $3\frac{1}{5}$ units to the right of the y-axis? Name the point, and give its coordinate pair.

    d. Which point is located $1\frac{1}{5}$ units from the y-axis?

    e. Which point is located $\frac{2}{5}$ unit along the x-axis?

    f. Give the coordinate pair for each of the following points.

        T: _____ U: _____ S: _____ K: _____

    g. Name the points located at the following coordinates.

        $(\frac{3}{5}, \frac{3}{5})$ _____ $(3\frac{2}{5}, 0)$ _____ $(2\frac{1}{5}, 3)$ _____ $(0, 2\frac{3}{5})$ _____

    h. Plot a point whose x- and y-coordinates are equal. Label your point E.

    i. What is the name for the point on the plane where the two axes intersect? _____

        Give the coordinates for this point. (_____ , _____)

    j. Plot the following points.

        A: $(1\frac{1}{5}, 1)$ B: $(\frac{1}{5}, 3)$ C: $(2\frac{4}{5}, 2\frac{2}{5})$ D: $(1\frac{1}{5}, 0)$

    k. What is the distance between L and N, or LN?

Lesson 3:   Name points using coordinate pairs, and use the coordinate pairs to plot points.

©2015 Great Minds eureka-math.org
G5-M6-SE-BK3-1.3.1-02.2016

l. What is the distance of MQ?

m. Would RM be greater than, less than, or equal to $LN + MQ$?

n. Leslie was explaining how to plot points on the coordinate plane to a new student, but she left off some important information. Correct her explanation so that it is complete.

"All you have to do is read the coordinates; for example, if it says (4, 7), count four, then seven, and put a point where the two grid lines intersect."

Lesson 3:   Name points using coordinate pairs, and use the coordinate pairs to plot points.

©2015 Great Minds eureka-math.org
G5-M6-SE-BK3-1.3.1-02.2016

19

This page intentionally left blank

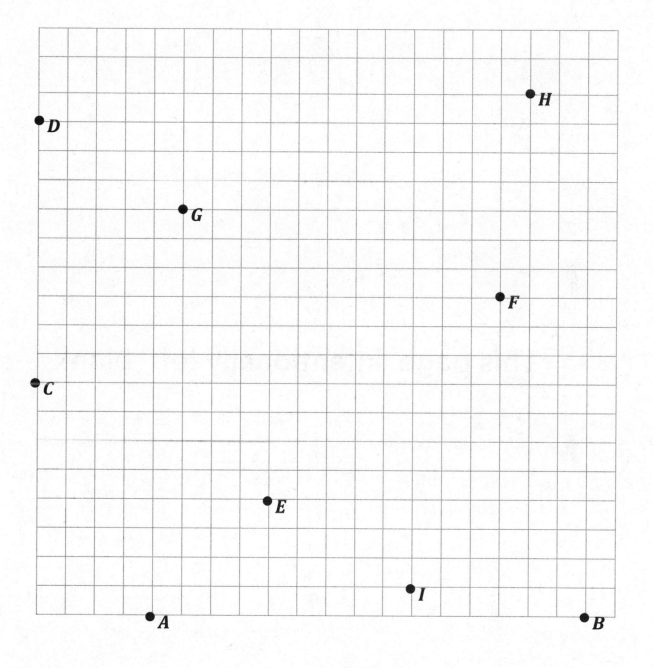

unlabeled coordinate plane

Lesson 3: Name points using coordinate pairs, and use the coordinate pairs to plot points.

This page intentionally left blank

Battleship Rules

Goal: To sink all of your opponent's ships by correctly guessing their coordinates.

Materials

- 1 grid sheet (per person/per game)
- Red crayon/marker for hits
- Black crayon/marker for misses
- Folder to place between players

Ships

- Each player must mark 5 ships on the grid.
    - Aircraft carrier—plot 5 points.
    - Battleship—plot 4 points.
    - Cruiser—plot 3 points.
    - Submarine—plot 3 points.
    - Patrol boat—plot 2 points.

Setup

- With your opponent, choose a unit length and fractional unit for the coordinate plane.
- Label the chosen units on both grid sheets.
- Secretly select locations for each of the 5 ships on your My Ships grid.
    - All ships must be placed horizontally or vertically on the coordinate plane.
    - Ships can touch each other, but they may not occupy the same coordinate.

Play

- Players take turns firing one shot to attack enemy ships.
- On your turn, call out the coordinates of your attacking shot. Record the coordinates of each attack shot.
- Your opponent checks his/her My Ships grid. If that coordinate is unoccupied, your opponent says, "Miss." If you named a coordinate occupied by a ship, your opponent says, "Hit."
- Mark each attempted shot on your Enemy Ships grid. Mark a black ✖ on the coordinate if your opponent says, "Miss." Mark a red ✓ on the coordinate if your opponent says, "Hit."
- On your opponent's turn, if he/she hits one of your ships, mark a red ✓ on that coordinate of your My Ships grid. When one of your ships has every coordinate marked with a ✓, say, "You've sunk my [name of ship]."

Victory

- The first player to sink all (or the most) opposing ships, wins.

Lesson 4:   Name points using coordinate pairs, and use the coordinate pairs to plot points.

My Ships

- Draw a red ✓ over any coordinate your opponent hits.
- Once all of the coordinates of any ship have been hit, say, "You've sunk my [name of ship]."

Aircraft carrier—5 points
Battleship—4 points
Cruiser—3 points
Submarine—3 points
Patrol boat—2 points

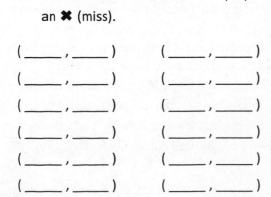

Enemy Ships

- Draw a black ✖ on the coordinate if your opponent says, "Miss."
- Draw a red ✓ on the coordinate if your opponent says, "Hit."
- Draw a circle around the coordinates of a sunken ship.

Attack Shots

- Record the coordinates of each shot below and whether it was a ✓ (hit) or an ✖ (miss).

(_____ , _____)   (_____ , _____)

(_____ , _____)   (_____ , _____)

(_____ , _____)   (_____ , _____)

(_____ , _____)   (_____ , _____)

(_____ , _____)   (_____ , _____)

(_____ , _____)   (_____ , _____)

(_____ , _____)   (_____ , _____)

(_____ , _____)   (_____ , _____)

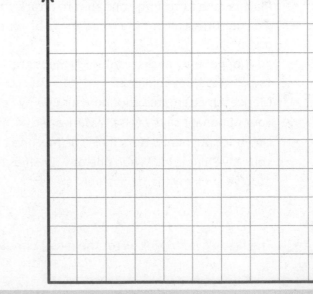

Lesson 4:   Name points using coordinate pairs, and use the coordinate pairs to plot points.

©2015 Great Minds eureka-math.org
G5-M6-SE-BK3-1.3.1-02.2016

Name _____   Date _____

Your homework is to play at least one game of Battleship with a friend or family member. You can use the directions from class to teach your opponent. You and your opponent should record your guesses, hits, and misses on the sheet as you did in class.

When you have finished your game, answer these questions.

1. When you guess a point that is a hit, how do you decide which points to guess next?

2. How could you change the coordinate plane to make the game easier or more challenging?

3. Which strategies worked best for you when playing this game?

This page intentionally left blank

Name _____   Date _____

1. Use the coordinate plane to the right to answer the
    following questions.

    a. Use a straightedge to construct a line that goes
        through points A and B. Label the line e.

    b. Line e is parallel to the _____-axis and is
        perpendicular to the _____-axis.

    c. Plot two more points on line e. Name them
        C and D.

    d. Give the coordinates of each point below.

        A: _____   B: _____

        C: _____   D: _____

    e. What do all of the points of line e have in common?

    f. Give the coordinates of another point that would fall on line e with an x-coordinate greater than 15.

Lesson 5:   Investigate patterns in vertical and horizontal lines, and interpret
points on the plane as distances from the axes.

27

2. Plot the following points on the coordinate plane to the right.

P: $(1\frac{1}{2}, \frac{1}{2})$ Q: $(1\frac{1}{2}, 2\frac{1}{2})$

R: $(1\frac{1}{2}, 1\frac{1}{4})$ S: $(1\frac{1}{2}, \frac{3}{4})$

a. Use a straightedge to draw a line to connect these points. Label the line h.

b. In line h, $x =$ _____ for all values of y.

c. Circle the correct word.

Line h is   *parallel*   *perpendicular*   to the x-axis.

Line h is   *parallel*   *perpendicular*   to the y-axis.

d. What pattern occurs in the coordinate pairs that let you know that line h is vertical?

3. For each pair of points below, think about the line that joins them. For which pairs is the line parallel to the x-axis? Circle your answer(s). Without plotting them, explain how you know.

a. (1.4, 2.2) and (4.1, 2.4)   b. (3, 9) and (8, 9)   c. $(1\frac{1}{4}, 2)$ and $(1\frac{1}{4}, 8)$

4. For each pair of points below, think about the line that joins them. For which pairs is the line parallel to the y-axis? Circle your answer(s). Then, give 2 other coordinate pairs that would also fall on this line.

a. (4, 12) and (6, 12)   b. $(\frac{3}{5}, 2\frac{3}{5})$ and $(\frac{1}{5}, 3\frac{1}{5})$ c. (0.8, 1.9) and (0.8, 2.3)

Lesson 5:   Investigate patterns in vertical and horizontal lines, and interpret points on the plane as distances from the axes.

5. Write the coordinate pairs of 3 points that can be connected to construct a line that is $5\frac{1}{2}$ units to the right of and parallel to the y-axis.

   a. _____   b. _____   c. _____

6. Write the coordinate pairs of 3 points that lie on the x-axis.

   a. _____   b. _____   c. _____

7. Adam and Janice are playing Battleship. Presented in the table is a record of Adam's guesses so far.
   He has hit Janice's battleship using these coordinate pairs. What should he guess next? How do you know? Explain using words and pictures.

(3, 11)	hit
(2, 11)	miss
(3, 10)	hit
(4, 11)	miss
(3, 9)	miss

Lesson 5:   Investigate patterns in vertical and horizontal lines, and interpret points on the plane as distances from the axes.

©2015 Great Minds eureka-math.org
G5-M6-SE-BK3-1.3.1-02.2016

29

This page intentionally left blank

Name _____   Date _____

1. Use the coordinate plane to answer the questions.

    a. Use a straightedge to construct a line that goes
        through points A and B. Label the line g.

    b. Line g is parallel to the _____-axis and is
        perpendicular to the _____-axis.

    c. Draw two more points on line g. Name them
        C and D.

    d. Give the coordinates of each point below.

        A: _____   B: _____

        C: _____   D: _____

    e. What do all of the points on line g have in common?

    f. Give the coordinates of another point that falls on line g with an x-coordinate greater than 25.

   Lesson 5:   Investigate patterns in vertical and horizontal lines, and interpret   31
                  points on the plane as distances from the axes.

©2015 Great Minds eureka-math.org
G5-M6-SE-BK3-1.3.1-02.2016

2. Plot the following points on the coordinate plane to the right.

   $H: (\frac{3}{4}, 3)$ $I: (\frac{3}{4}, 2\frac{1}{4})$

   $J: (\frac{3}{4}, \frac{1}{2})$ $K: (\frac{3}{4}, 1\frac{3}{4})$

   a. Use a straightedge to draw a line to connect these points. Label the line f.

   b. In line f, $x =$ _____ for all values of y.

   c. Circle the correct word:

      Line f is *parallel*   *perpendicular* to the x-axis.

      Line f is *parallel*   *perpendicular* to the y-axis.

   d. What pattern occurs in the coordinate pairs that make line f vertical?

3. For each pair of points below, think about the line that joins them. For which pairs is the line parallel to the x-axis? Circle your answer(s). Without plotting them, explain how you know.

   a. (3.2, 7) and (5, 7)   b. (8, 8.4) and (8, 8.8)   c. $(6\frac{1}{2}, 12)$ and (6.2, 11)

4. For each pair of points below, think about the line that joins them. For which pairs is the line parallel to the y-axis? Circle your answer(s). Then, give 2 other coordinate pairs that would also fall on this line.

   a. (3.2, 8.5) and (3.22, 24)   b. $(13\frac{1}{3}, 4\frac{2}{3})$ and $(13\frac{1}{3}, 7)$ c. (2.9, 5.4) and (7.2, 5.4)

Lesson 5: Investigate patterns in vertical and horizontal lines, and interpret points on the plane as distances from the axes.

©2015 Great Minds eureka-math.org
G5-M6-SE-BK3-1.3.1-02.2016

5. Write the coordinate pairs of 3 points that can be connected to construct a line that is $5\frac{1}{2}$ units to the right of and parallel to the y-axis.

   a. _____   b. _____   c. _____

6. Write the coordinate pairs of 3 points that lie on the y-axis.

   a. _____   b. _____   c. _____

7. Leslie and Peggy are playing Battleship on axes labeled in halves. Presented in the table is a record of Peggy's guesses so far. What should she guess next? How do you know? Explain using words and pictures.

(5, 5)	miss
(4, 5)	hit
$(3\frac{1}{2}, 5)$	miss
$(4\frac{1}{2}, 5)$	miss

Lesson 5: Investigate patterns in vertical and horizontal lines, and interpret points on the plane as distances from the axes.

©2015 Great Minds eureka-math.org
G5-M6-SE-BK3-1.3.1-02.2016

33

This page intentionally left blank

Point	x	y	(x, y)
H			
I			
J			
K			
L			

a.

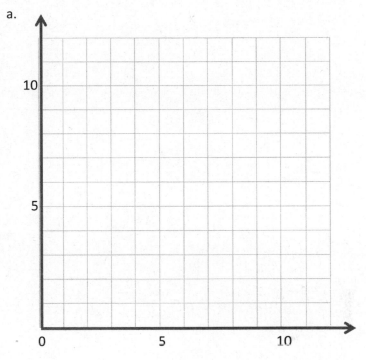

b.

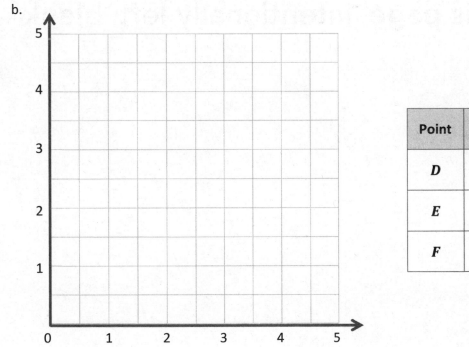

Point	x	y	(x, y)
D	$2\frac{1}{2}$	0	$(2\frac{1}{2}, 0)$
E	$2\frac{1}{2}$	2	$(2\frac{1}{2}, 2)$
F	$2\frac{1}{2}$	4	$(2\frac{1}{2}, 4)$

coordinate plane practice

Lesson 5:   Investigate patterns in vertical and horizontal lines, and interpret
              points on the plane as distances from the axes.

35

©2015 Great Minds eureka-math.org
G5-M6-SE-BK3-1.3.1-02.2016

This page intentionally left blank

Name _____ Date _____

1. Plot the following points, and label them on the coordinate plane.

   A: (0.3, 0.1)   B: (0.3, 0.7)

   C: (0.2, 0.9)   D: (0.4, 0.9)

   a. Use a straightedge to construct line segments $\overline{AB}$ and $\overline{CD}$.

   b. Line segment _____ is parallel to the x-axis and is perpendicular to the y-axis.

   c. Line segment _____ is parallel to the y-axis and is perpendicular to the x-axis.

   d. Plot a point on line segment $\overline{AB}$ that is not at the endpoints, and name it U. Write the coordinates.
      U (_____ , _____)

   e. Plot a point on line segment $\overline{CD}$, and name it V. Write the coordinates. V (_____ , _____)

Lesson 6: Investigate patterns in vertical and horizontal lines, and interpret points on the plane as distances from the axes.

©2015 Great Minds eureka-math.org
G5-M6-SE-BK3-1.3.1-02.2016

37

2. Construct line f such that the y-coordinate of every point is $3\frac{1}{2}$, and construct line g such that the x-coordinate of every point is $4\frac{1}{2}$.

    a. Line f is _____ units from the x-axis.

    b. Give the coordinates of the point on line f that is $\frac{1}{2}$ unit from the y-axis. _____

    c. With a blue pencil, shade the portion of the grid that is less than $3\frac{1}{2}$ units from the x-axis.

    d. Line g is _____ units from the y-axis.

    e. Give the coordinates of the point on line g that is 5 units from the x-axis. _____

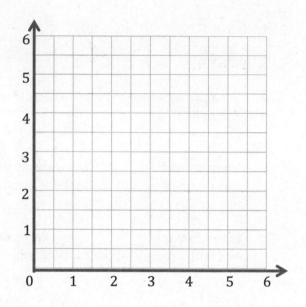

    f. With a red pencil, shade the portion of the grid that is more than $4\frac{1}{2}$ units from the y-axis.

Lesson 6:   Investigate patterns in vertical and horizontal lines, and interpret points on the plane as distances from the axes.

©2015 Great Minds eureka-math.org
G5-M6-SE-BK3-1.3.1-02.2016

3. Complete the following tasks on the plane below.

   a. Construct a line m that is perpendicular to the x-axis and 3.2 units from the y-axis.

   b. Construct a line a that is 0.8 unit from the x-axis.

   c. Construct a line t that is parallel to line m and is halfway between line m and the y-axis.

   d. Construct a line h that is perpendicular to line t and passes through the point (1.2, 2.4).

   e. Using a blue pencil, shade the region that contains points that are more than 1.6 units and less than 3.2 units from the y-axis.

   f. Using a red pencil, shade the region that contains points that are more than 0.8 unit and less than 2.4 units from the x-axis.

   g. Give the coordinates of a point that lies in the double-shaded region.

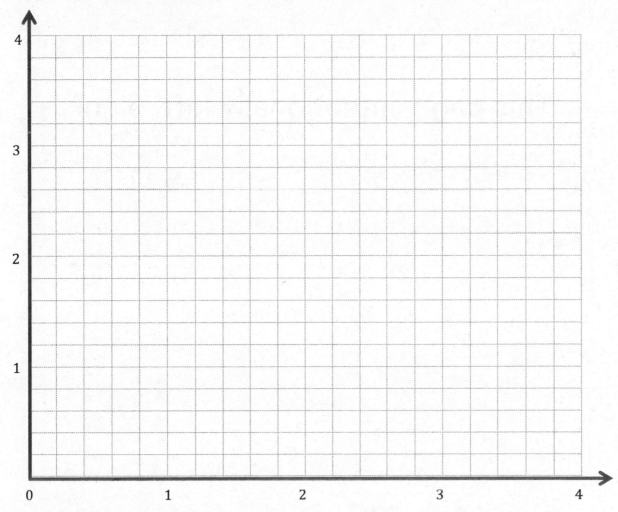

Lesson 6: Investigate patterns in vertical and horizontal lines, and interpret points on the plane as distances from the axes.

©2015 Great Minds eureka-math.org
G5-M6-SE-BK3-1.3.1-02.2016

39

This page intentionally left blank

Name _____ Date _____

1. Plot and label the following points on the coordinate plane.

    C: (0.4, 0.4)   A: (1.1, 0.4)   S: (0.9, 0.5)   T: (0.9, 1.1)

    a. Use a straightedge to construct line segments $\overline{CA}$ and $\overline{ST}$.

    b. Name the line segment that is perpendicular to the x-axis and parallel to the y-axis.

       _____

    c. Name the line segment that is parallel to the x-axis and perpendicular to the y-axis.

       _____

    d. Plot a point on $\overline{CA}$, and name it E. Plot a point on line segment $\overline{ST}$, and name it R.

    e. Write the coordinates of points E and R.

       E (____ , ____)   R (____ , ____)

Lesson 6:   Investigate patterns in vertical and horizontal lines, and interpret points on the plane as distances from the axes.

41

©2015 Great Minds eureka-math.org
G5-M6-SE-BK3-1.3.1-02.2016

2. Construct line m such that the y-coordinate of every point is $1\frac{1}{2}$, and construct line n such that the x-coordinate of every point is $5\frac{1}{2}$.

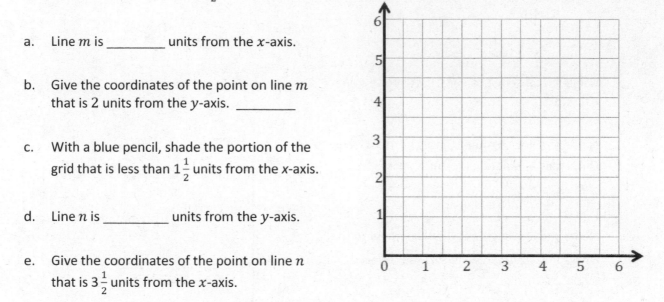

   a.   Line m is _____ units from the x-axis.

   b.   Give the coordinates of the point on line m that is 2 units from the y-axis. _____

   c.   With a blue pencil, shade the portion of the grid that is less than $1\frac{1}{2}$ units from the x-axis.

   d.   Line n is _____ units from the y-axis.

   e.   Give the coordinates of the point on line n that is $3\frac{1}{2}$ units from the x-axis.

   f.   With a red pencil, shade the portion of the grid that is less than $5\frac{1}{2}$ units from the y-axis.

Lesson 6:   Investigate patterns in vertical and horizontal lines, and interpret points on the plane as distances from the axes.

©2015 Great Minds eureka-math.org
G5-M6-SE-BK3-1.3.1-02.2016

3. Construct and label lines e, r, s, and o on the plane below.

   a. Line e is 3.75 units above the x-axis.

   b. Line r is 2.5 units from the y-axis.

   c. Line s is parallel to line e but 0.75 farther from the x-axis.

   d. Line o is perpendicular to lines s and e and passes through the point $(3\frac{1}{4}, 3\frac{1}{4})$.

4. Complete the following tasks on the plane.

   a. Using a blue pencil, shade the region that contains points that are more than $2\frac{1}{2}$ units and less than $3\frac{1}{4}$ units from the y-axis.

   b. Using a red pencil, shade the region that contains points that are more than $3\frac{3}{4}$ units and less than $4\frac{1}{2}$ units from the x-axis.

   c. Plot a point that lies in the double-shaded region, and label its coordinates.

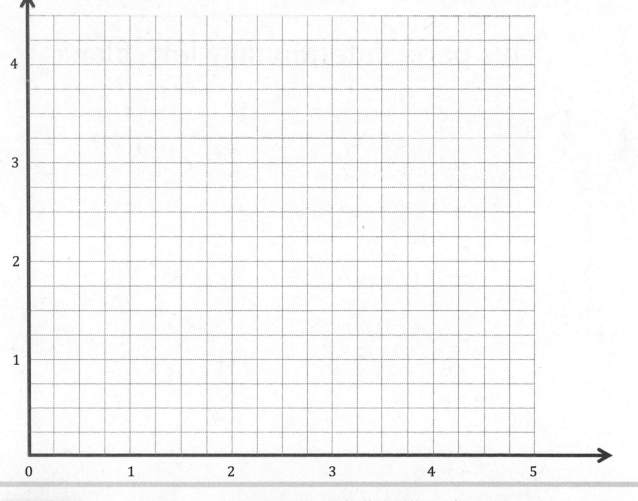

Lesson 6:   Investigate patterns in vertical and horizontal lines, and interpret
              points on the plane as distances from the axes.

43

©2015 Great Minds eureka-math.org
G5-M6-SE-BK3-1.3.1-02.2016

This page intentionally left blank

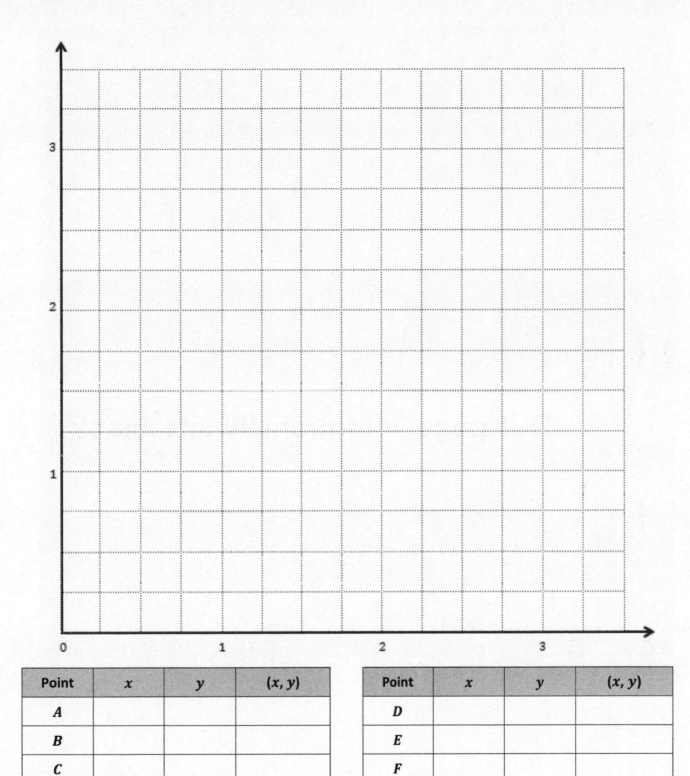

Point	x	y	(x, y)
A			
B			
C			

Point	x	y	(x, y)
D			
E			
F			

coordinate plane

This page intentionally left blank

Name _____   Date _____

1. Complete the chart. Then, plot the points on the coordinate plane below.

x	y	(x, y)
0	1	(0, 1)
2	3	
4	5	
6	7	

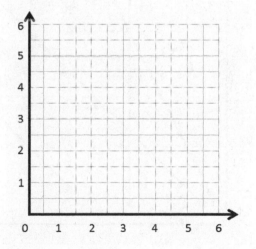

a. Use a straightedge to draw a line connecting these points.

b. Write a rule showing the relationship between the x- and y-coordinates of points on the line.

c. Name 2 other points that are on this line. _____   _____

2. Complete the chart. Then, plot the points on the coordinate plane below.

x	y	(x, y)
$\frac{1}{2}$	1	
1	2	
$1\frac{1}{2}$	3	
2	4	

a. Use a straightedge to draw a line connecting these points.

b. Write a rule showing the relationship between the x- and y-coordinates.

c. Name 2 other points that are on this line. _____   _____

Lesson 7:   Plot points, use them to draw lines in the plane, and describe patterns
             within the coordinate pairs.

3. Use the coordinate plane below to answer the following questions.

a. Give the coordinates for 3 points that are on line a. _____ _____ _____

b. Write a rule that describes the relationship between the x- and y-coordinates for the points on line a.

Lesson 7:   Plot points, use them to draw lines in the plane, and describe patterns within the coordinate pairs.

©2015 Great Minds eureka-math.org
G5-M6-SE-BK3-1.3.1-02.2016

c. What do you notice about the y-coordinates of every point on line b?

d. Fill in the missing coordinates for points on line d.

(12, _____)   (6, _____)   (_____, 24)   (28, _____)   (_____, 28)

e. For any point on line c, the x-coordinate is _____.

f. Each of the points lies on at least 1 of the lines shown in the plane on the previous page. Identify a line that contains each of the following points.

i. (7, 7) ___a___   ii. (14, 8) _____   iii. (5, 10) _____

iv. (0, 17) _____   v. (15.3, 9.3) _____   vi. (20, 40) _____

Lesson 7:   Plot points, use them to draw lines in the plane, and describe patterns
            within the coordinate pairs.

©2015 Great Minds eureka-math.org
G5-M6-SE-BK3-1.3.1-02.2016

49

This page intentionally left blank

Name _____   Date _____

1. Complete the chart. Then, plot the points on the coordinate plane.

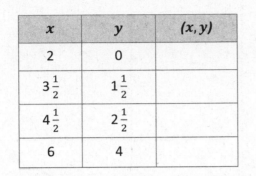

x	y	(x, y)
2	0	
$3\frac{1}{2}$	$1\frac{1}{2}$	
$4\frac{1}{2}$	$2\frac{1}{2}$	
6	4	

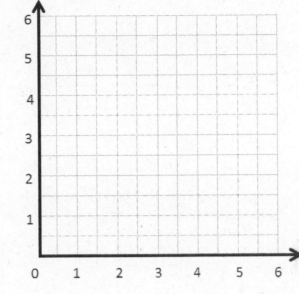

a. Use a straightedge to draw a line connecting these points.

b. Write a rule showing the relationship between the x- and y-coordinates of points on this line.

c. Name two other points that are also on this line. _____   _____

2. Complete the chart. Then, plot the points on the coordinate plane.

x	y	(x, y)
0	0	
$\frac{1}{4}$	$\frac{3}{4}$	
$\frac{1}{2}$	$1\frac{1}{2}$	
1	3	

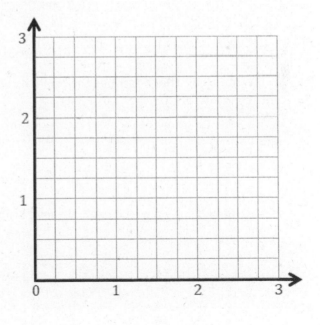

a. Use a straightedge to draw a line connecting these points.

b. Write a rule showing the relationship between the x- and y-coordinates for points on the line.

c. Name two other points that are also on this line. _____   _____

3. Use the coordinate plane to answer the following questions.

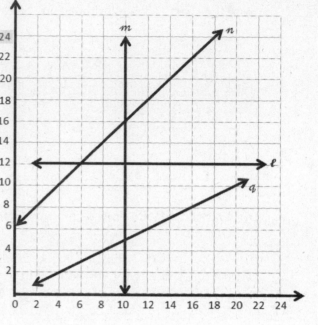

a. For any point on line *m*, the *x*-coordinate is

_____.

b. Give the coordinates for 3 points that are on line *n*.

c. Write a rule that describes the relationship between the *x*- and *y*-coordinates on line *n*.

d. Give the coordinates for 3 points that are on line *q*.

e. Write a rule that describes the relationship between the *x*- and *y*-coordinates on line *q*.

f. Identify a line on which each of these points lie.

   i. (10, 3.2) _____   ii. (12.4, 18.4) _____

   iii. (6.45, 12) _____   iv. (14, 7) _____

Lesson 7:   Plot points, use them to draw lines in the plane, and describe patterns
                    within the coordinate pairs.

©2015 Great Minds eureka-math.org
G5-M6-SE-BK3-1.3.1-02.2016

EUREKA
MATH™

Name _____   Date _____

1.

a.

Point	x	y	(x, y)
A	0	0	(0, 0)
B	1	1	(1, 1)
C	2	2	(2, 2)
D	3	3	(3, 3)

b.

Point	x	y	(x, y)
G	0	3	(0, 3)
H	$\frac{1}{2}$	$3\frac{1}{2}$	$(\frac{1}{2}, 3\frac{1}{2})$
I	1	4	(1, 4)
J	$1\frac{1}{2}$	$4\frac{1}{2}$	$(1\frac{1}{2}, 4\frac{1}{2})$

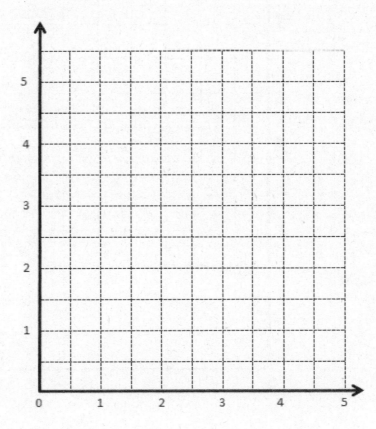

coordinate plane

EUREKA MATH

Lesson 7:   Plot points, use them to draw lines in the plane, and describe patterns within the coordinate pairs.

©2015 Great Minds eureka-math.org
G5-M6-SE-BK3-1.3.1-02.2016

53

2.

a.

Point	(x, y)
L	(0, 3)
M	(2, 3)
N	(4, 3)

b.

Point	(x, y)
O	(0, 0)
P	(1, 2)
Q	(2, 4)

c.

Point	(x, y)
R	$(1, \frac{1}{2})$
S	$(2, 1\frac{1}{2})$
T	$(3, 2\frac{1}{2})$

d.

Point	(x, y)
U	(1, 3)
V	(2, 6)
W	(3, 9)

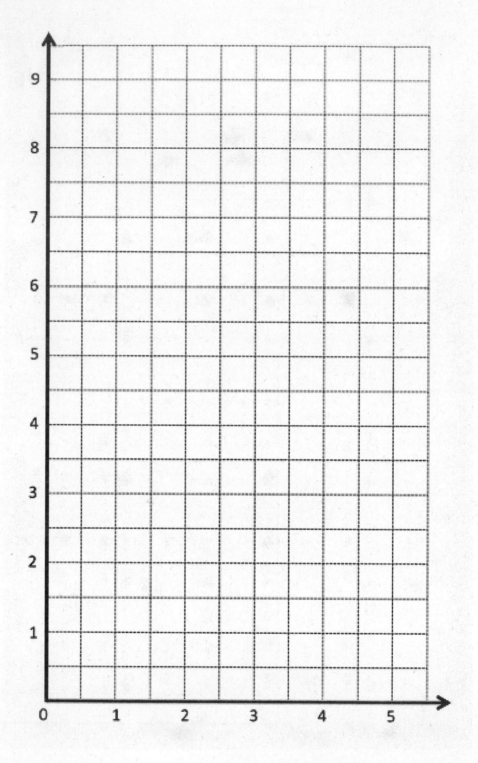

coordinate plane

Lesson 7:   Plot points, use them to draw lines in the plane, and describe patterns within the coordinate pairs.

Name _____   Date _____

1. Create a table of 3 values for x and y such that each y-coordinate is 3 more than the corresponding x-coordinate.

x	y	(x, y)

a. Plot each point on the coordinate plane.

b. Use a straightedge to draw a line connecting these points.

c. Give the coordinates of 2 other points that fall on this line with x-coordinates greater than 12.

(_____ , _____) and (_____ , _____)

2. Create a table of 3 values for x and y such that each y-coordinate is 3 times as much as its corresponding x-coordinate.

x	y	(x, y)

a. Plot each point on the coordinate plane.

b. Use a straightedge to draw a line connecting these points.

c. Give the coordinates of 2 other points that fall on this line with y-coordinates greater than 25.

(_____ , _____) and (_____ , _____)

Lesson 8:   Generate a number pattern from a given rule, and plot the points.

©2015 Great Minds eureka-math.org
G5-M6-SE-BK3-1.3.1-02.2016

3. Create a table of 5 values for x and y such that each y-coordinate is 1 more than 3 times as much as its corresponding x value.

x	y	(x, y)

a. Plot each point on the coordinate plane.

b. Use a straightedge to draw a line connecting these points.

c. Give the coordinates of 2 other points that would fall on this line whose x-coordinates are greater than 12.
   (_____ , _____) and (_____ , _____)

Lesson 8: Generate a number pattern from a given rule, and plot the points.

57

©2015 Great Minds eureka-math.org
G5-M6-SE-BK3-1.3.1-02.2016

4. Use the coordinate plane below to complete the following tasks.

a. Graph the lines on the plane.

line ℓ: x is equal to y

	x	y	(x, y)
A			
B			
C			

line m: y is 1 more than x

	x	y	(x, y)
G			
H			
I			

line n: y is 1 more than twice x

	x	y	(x, y)
S			
T			
U			

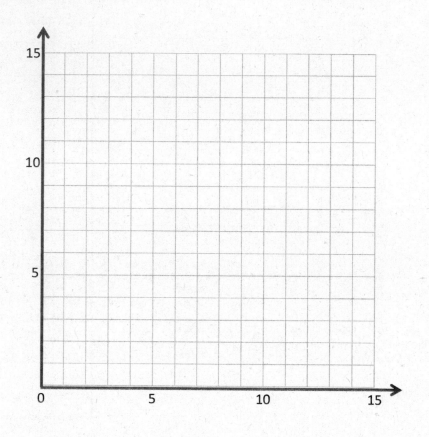

b. Which two lines intersect? Give the coordinates of their intersection.

c. Which two lines are parallel?

d. Give the rule for another line that would be parallel to the lines you listed in Problem 4(c).

Lesson 8: Generate a number pattern from a given rule, and plot the points.

©2015 Great Minds eureka-math.org
G5-M6-SE-BK3-1.3.1-02.2016

Name _____   Date _____

1.   Complete this table such that each y-coordinate is 4 more than the corresponding x-coordinate.

x	y	(x, y)

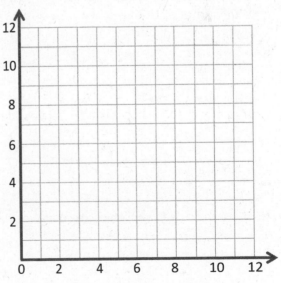

   a.   Plot each point on the coordinate plane.

   b.   Use a straightedge to construct a line connecting these points.

   c.   Give the coordinates of 2 other points that fall on this line with x-coordinates greater than 18.

   (_____ , _____) and (_____ , _____)

2.   Complete this table such that each y-coordinate is 2 times as much as its corresponding x-coordinate.

x	y	(x, y)

   a.   Plot each point on the coordinate plane.

   b.   Use a straightedge to draw a line connecting these points.

   c.   Give the coordinates of 2 other points that fall on this line with y-coordinates greater than 25.

   (_____ , _____) and (_____ , _____)

EUREKA MATH™

©2015 Great Minds eureka-math.org
G5-M6-SE-BK3-1.3.1-02.2016

3. Use the coordinate plane below to complete the following tasks.

    a. Graph these lines on the plane.

        line ℓ: x is equal to y

	x	y	(x, y)
A			
B			
C			

        line m: y is 1 less than x

	x	y	(x, y)
G			
H			
I			

        line n: y is 1 less than twice x

	x	y	(x, y)
S			
T			
U			

    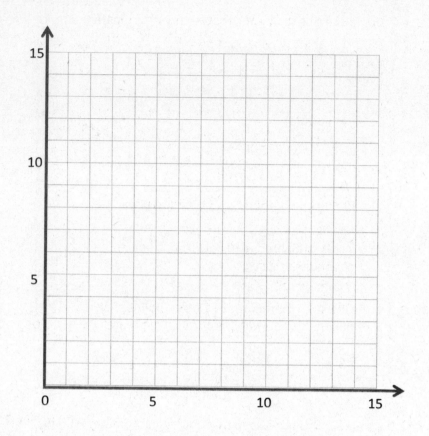

    b. Do any of these lines intersect? If yes, identify which ones, and give the coordinates of their intersection.

    c. Are any of these lines parallel? If yes, identify which ones.

    d. Give the rule for another line that would be parallel to the lines you listed in Problem 3(c).

Lesson 8:   Generate a number pattern from a given rule, and plot the points.

©2015 Great Minds eureka-math.org
G5-M6-SE-BK3-1.3.1-02.2016

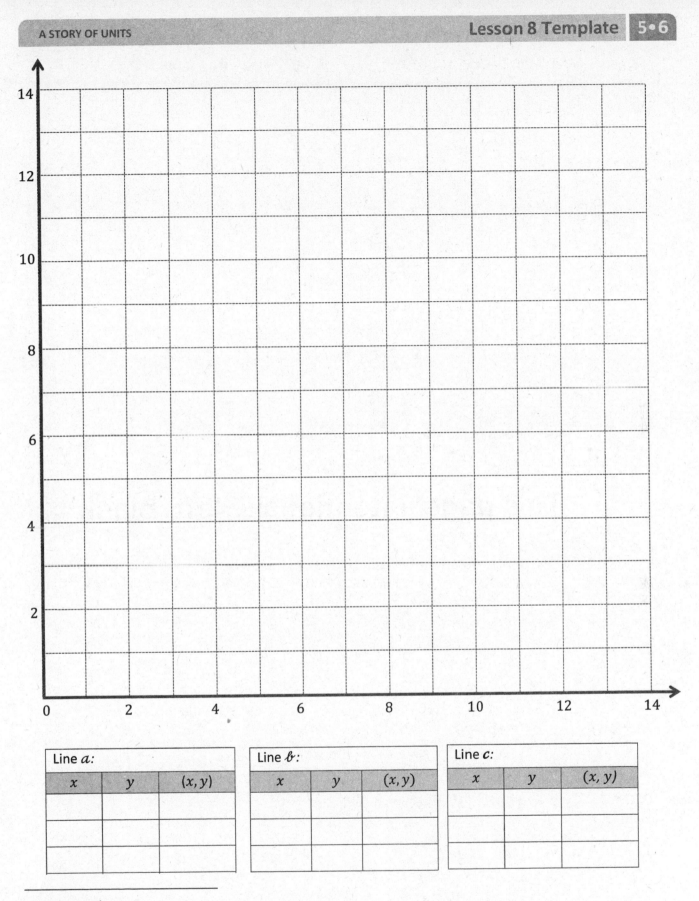

Line a:		
x	y	(x, y)

Line b:		
x	y	(x, y)

Line c:		
x	y	(x, y)

coordinate plane

This page intentionally left blank

Name _____ Date _____

1. Complete the table for the given rules.

   Line *a*

   Rule: *y is 1 more than x*

x	y	(x, y)
1		
5		
9		
13		

   Line *b*

   Rule: *y is 4 more than x*

x	y	(x, y)
0		
5		
8		
11		

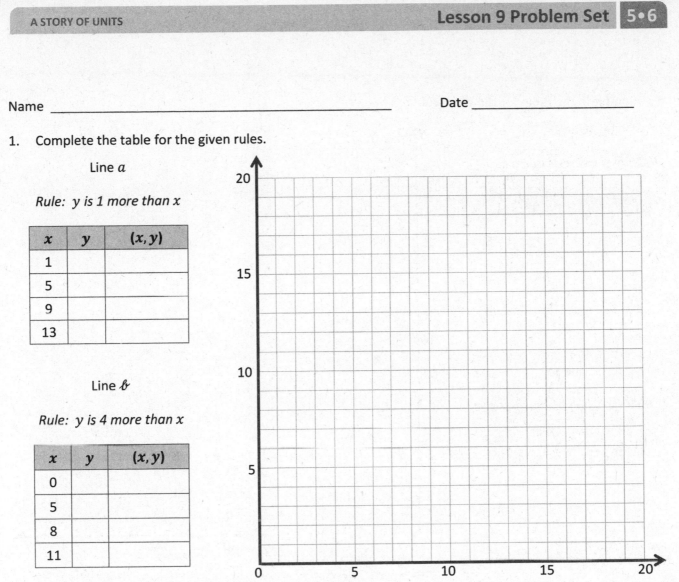

   a. Construct each line on the coordinate plane above.

   b. Compare and contrast these lines.

   c. Based on the patterns you see, predict what line *c*, whose rule is *y is 7 more than x*, would look like. Draw your prediction on the plane above.

Lesson 9: Generate two number patterns from given rules, plot the points, and analyze the patterns.

63

©2015 Great Minds eureka-math.org
G5-M6-SE-BK3-1.3.1-02.2016

2. Complete the table for the given rules.

Line *e*

Rule: *y is twice as much as x*

x	y	(x, y)
0		
2		
5		
9		

Line *f*

Rule: *y is half as much as x*

x	y	(x, y)
0		
6		
10		
20		

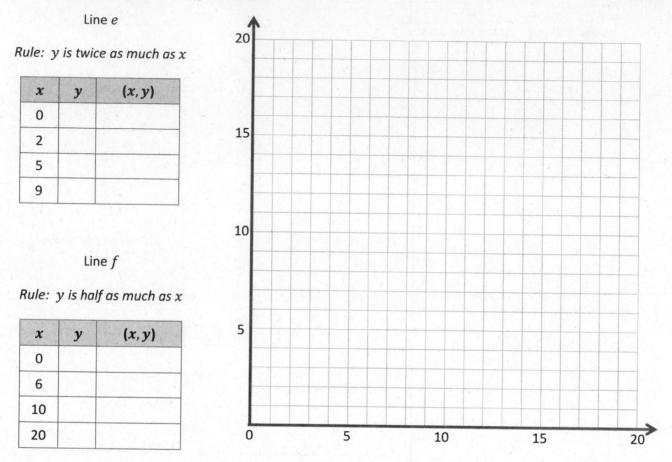

a. Construct each line on the coordinate plane above.

b. Compare and contrast these lines.

c. Based on the patterns you see, predict what line *g*, whose rule is *y is 4 times as much as x*, would look like. Draw your prediction in the plane above.

Lesson 9:   Generate two number patterns from given rules, plot the points, and analyze the patterns.

©2015 Great Minds eureka-math.org
G5-M6-SE-BK3-1.3.1-02.2016

Name _____ Date _____

1. Complete the table for the given rules.

Line *a*

Rule: *y is 1 less than x*

x	y	(x, y)
1		
4		
9		
16		

Line *b*

Rule: *y is 5 less than x*

x	y	(x, y)
5		
8		
14		
20		

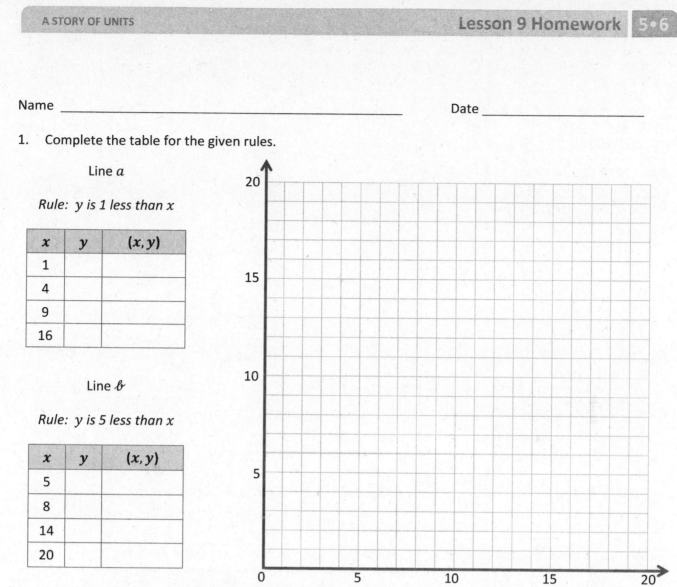

a. Construct each line on the coordinate plane.

b. Compare and contrast these lines.

c. Based on the patterns you see, predict what line *c*, whose rule is *y is 7 less than x*, would look like. Draw your prediction on the plane above.

Lesson 9: Generate two number patterns from given rules, plot the points, and analyze the patterns.

65

©2015 Great Minds eureka-math.org
G5-M6-SE-BK3-1.3.1-02.2016

2. Complete the table for the given rules.

Line e

Rule: y is 3 times as much as x

x	y	(x, y)
0		
1		
4		
6		

Line f

Rule: y is a third as much as x

x	y	(x, y)
0		
3		
9		
15		

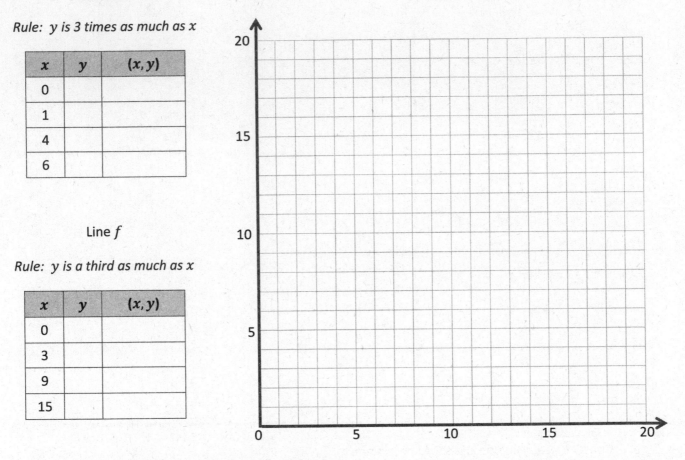

a. Construct each line on the coordinate plane.

b. Compare and contrast these lines.

c. Based on the patterns you see, predict what line g, whose rule is y is 4 times as much as x, and line h, whose rule is y is one-fourth as much as x, would look like. Draw your prediction in the plane above.

Lesson 9:   Generate two number patterns from given rules, plot the points, and
                            analyze the patterns.

©2015 Great Minds eureka-math.org
G5-M6-SE-BK3-1.3.1-02.2016

Line ℓ

Rule: *y is 2 more than x*

x	y	(x, y)
1		
5		
10		
15		

Line m

Rule: *y is 5 more than x*

x	y	(x, y)
0		
5		
10		
15		

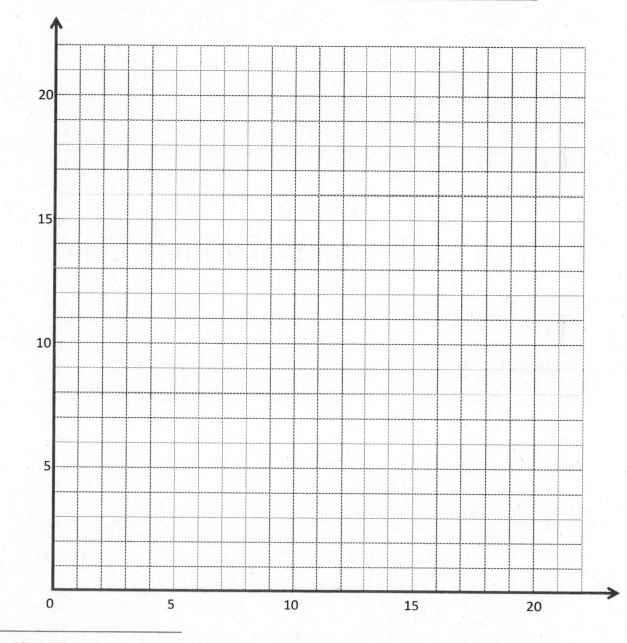

coordinate plane

Lesson 9: Generate two number patterns from given rules, plot the points, and analyze the patterns.

©2015 Great Minds eureka-math.org
G5-M6-SE-BK3-1.3.1-02.2016

67

This page intentionally left blank

Line p

Rule: y is x times 2

x	y	(x, y)

Line q

Rule: y is x times 3

x	y	(x, y)

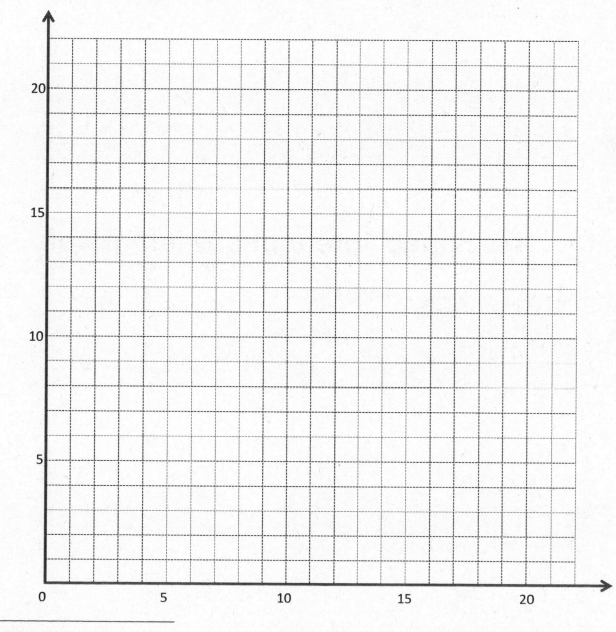

coordinate plane

Lesson 9: Generate two number patterns from given rules, plot the points, and
analyze the patterns.

69

©2015 Great Minds eureka-math.org
G5-M6-SE-BK3-1.3.1-02.2016

This page intentionally left blank

Name _____   Date _____

1. Use the coordinate plane below to complete the following tasks.

    a. Line p represents the rule *x and y are equal.*

    b. Construct a line, d, that is parallel to line p and contains point D.

    c. Name 3 coordinate pairs on line d.

    d. Identify a rule to describe line d.

    e. Construct a line, e, that is parallel to line p and contains point E.

    f. Name 3 points on line e.

    g. Identify a rule to describe line e.

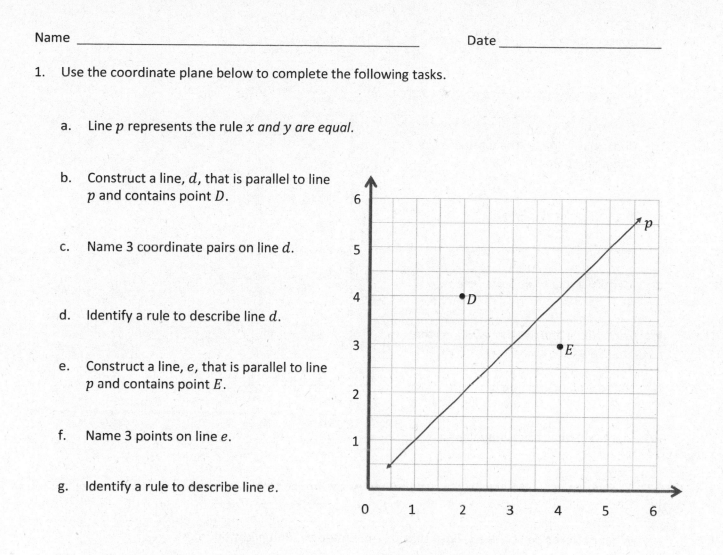

    h. Compare and contrast lines d and e in terms of their relationship to line p.

2. Write a rule for a fourth line that would be parallel to those above and would contain the point $(3\frac{1}{2}, 6)$. Explain how you know.

3. Use the coordinate plane below to complete the following tasks.

   a. Line p represents the rule x and y are equal.

   b. Construct a line, v, that contains the origin and point V.

   c. Name 3 points on line v.

   d. Identify a rule to describe line v.

   e. Construct a line, w, that contains the origin and point W.

   f. Name 3 points on line w.

   g. Identify a rule to describe line w.

   h. Compare and contrast lines v and w in terms of their relationship to line p.

   i. What patterns do you see in lines that are generated by multiplication rules?

4. Circle the rules that generate lines that are parallel to each other.

   add 5 to x   multiply x by $\frac{2}{3}$ x plus $\frac{1}{2}$ x times $1\frac{1}{2}$

Lesson 10:   Compare the lines and patterns generated by addition rules and multiplicative rules.

EUREKA MATH

Name _____   Date _____

1. Use the coordinate plane to complete the following tasks.

    a. Line p represents the rule *x and y are equal*.

    b. Construct a line, d, that is parallel to line p and contains point D.

    c. Name 3 coordinate pairs on line d.

    d. Identify a rule to describe line d.

    e. Construct a line, e, that is parallel to line p and contains point E.

    f. Name 3 points on line e.

    g. Identify a rule to describe line e.

    h. Compare and contrast lines d and e in terms of their relationship to line p.

2. Write a rule for a fourth line that would be parallel to those above and that would contain the point $(5\frac{1}{2}, 2)$. Explain how you know.

Lesson 10: Compare the lines and patterns generated by addition rules and multiplicative rules.

©2015 Great Minds eureka-math.org
G5-M6-SE-BK3-1.3.1-02.2016

73

3. Use the coordinate plane below to complete the following tasks.

    a. Line p represents the rule *x and y are equal*.

    b. Construct a line, v, that contains the origin and point V.

    c. Name 3 points on line v.

    d. Identify a rule to describe line v.

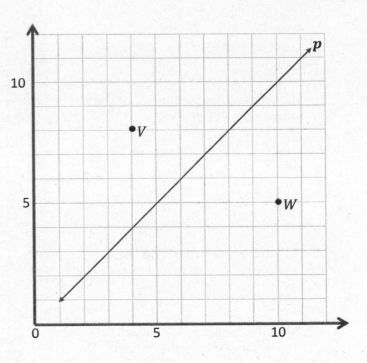

    e. Construct a line, w, that contains the origin and point W.

    f. Name 3 points on line w.

    g. Identify a rule to describe line w.

    h. Compare and contrast lines v and w in terms of their relationship to line p.

    i. What patterns do you see in lines that are generated by multiplication rules?

Lesson 10:   Compare the lines and patterns generated by addition rules and
               multiplicative rules.

©2015 Great Minds eureka-math.org
G5-M6-SE-BK3-1.3.1-02.2016

Line p

Line b

Line c

Line d

Rule: *y is 0 more than x*

Rule: _____

Rule: _____

Rule: _____

x	y	(x, y)
0		
5		
10		
15		

x	y	(x, y)
7		
10		
13		
18		

x	y	(x, y)
2		
4		
8		
11		

x	y	(x, y)
5		
7		
12		
15		

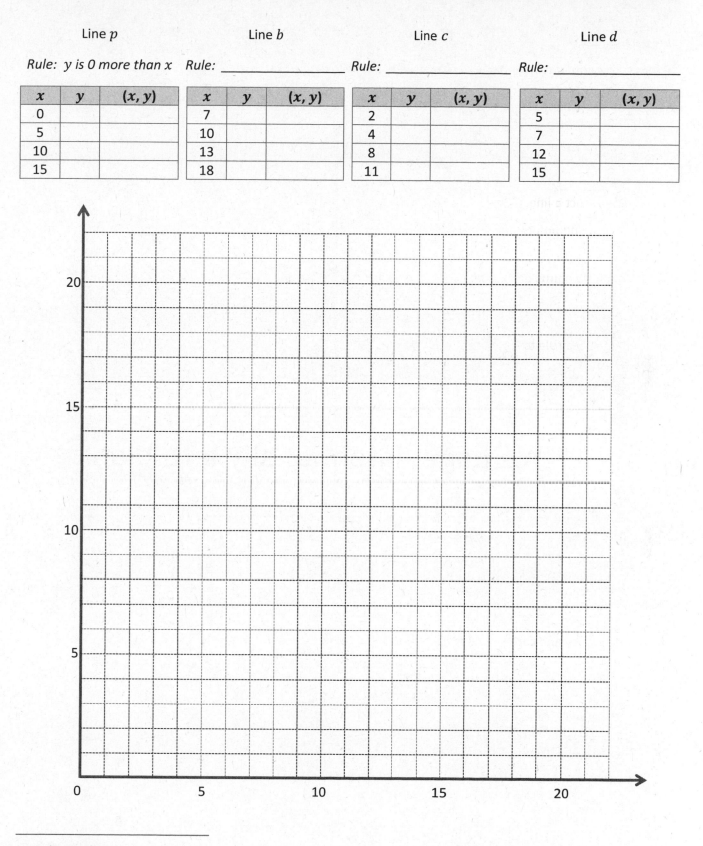

coordinate plane

EUREKA MATH™

Lesson 10: Compare the lines and patterns generated by addition rules and multiplicative rules.

©2015 Great Minds eureka-math.org
G5-M6-SE-BK3-1.3.1-02.2016

This page intentionally left blank

Line g Rule: _____

Line h Rule: _____

x	y	(x, y)
1		
2		
5		
7		

x	y	(x, y)
3		
6		
12		
15		

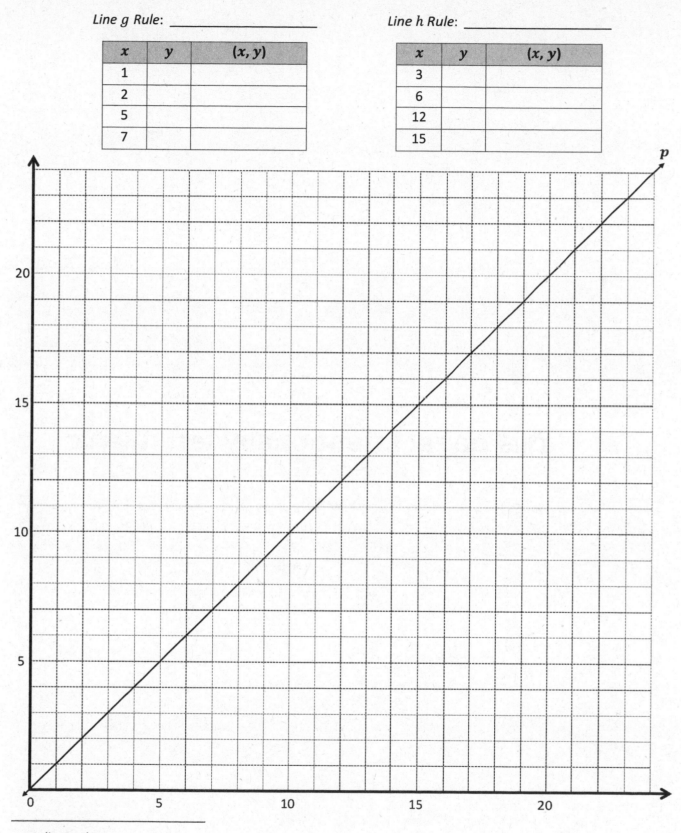

coordinate plane

Lesson 10: Compare the lines and patterns generated by addition rules and multiplicative rules.

©2015 Great Minds eureka-math.org
G5-M6-SE-BK3-1.3.1-02.2016

77

This page intentionally left blank

Name _____   Date _____

1. Complete the tables for the given rules.

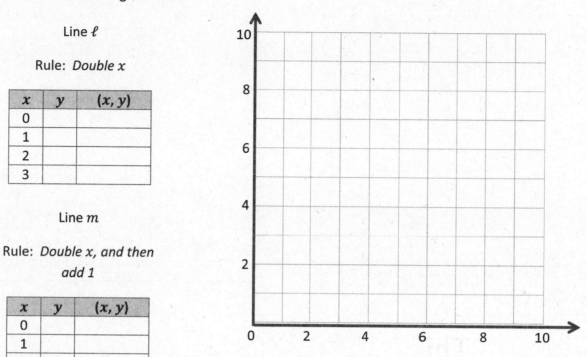

Line ℓ

Rule: *Double x*

x	y	(x, y)
0		
1		
2		
3		

Line m

Rule: *Double x, and then add 1*

x	y	(x, y)
0		
1		
2		
3		

a. Draw each line on the coordinate plane above.

b. Compare and contrast these lines.

c. Based on the patterns you see, predict what the line for the rule *double x, and then subtract 1* would look like. Draw the line on the plane above.

2. Circle the point(s) that the line for the rule *multiply x by $\frac{1}{3}$, and then add 1* would contain.

$(0, \frac{1}{3})$ $(2, 1\frac{2}{3})$ $(1\frac{1}{2}, 1\frac{1}{2})$ $(2\frac{1}{4}, 2\frac{1}{4})$

a. Explain how you know.

b. Give two other points that fall on this line.

3. Complete the tables for the given rules.

Line ℓ
Rule: *Halve x*

x	y	(x, y)
0		
1		
2		
3		

Line m
Rule: *Halve x, and then add $1\frac{1}{2}$*

x	y	(x, y)
0		
1		
2		
3		

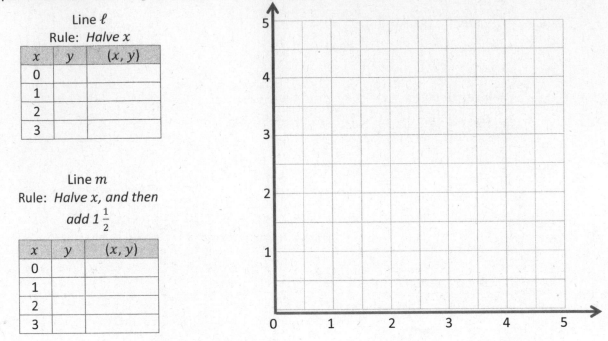

a. Draw each line on the coordinate plane above.

b. Compare and contrast these lines.

c. Based on the patterns you see, predict what the line for the rule *halve x, and then subtract 1* would look like. Draw the line on the plane above.

4. Circle the point(s) that the line for the rule *multiply x by $\frac{2}{3}$, and then subtract 1* would contain.

$(1\frac{1}{3}, \frac{1}{9})$ $(2, \frac{1}{3})$ $(1\frac{3}{2}, 1\frac{1}{2})$ $(3, 1)$

a. Explain how you know.

b. Give two other points that fall on this line.

Lesson 11:   Analyze number patterns created from mixed operations.

EUREKA MATH

©2015 Great Minds eureka-math.org
G5-M6-SE-BK3-1.3.1-02.2016

Name _____ Date _____

1. Complete the tables for the given rules.

Line ℓ

Rule: *Double x*

x	y	(x, y)
1	·	
2		
3		

Line m

Rule: *Double x, and then subtract 1*

x	y	(x, y)
1		
2		
3		

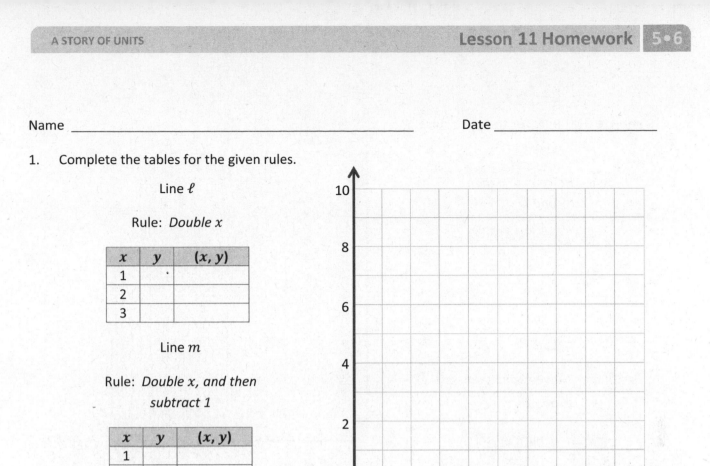

a. Draw each line on the coordinate plane above.

b. Compare and contrast these lines.

c. Based on the patterns you see, predict what the line for the rule *double x, and then add 1* would look like. Draw your prediction on the plane above.

2. Circle the point(s) that the line for the rule *multiply x by $\frac{1}{2}$, and then add 1* would contain.

$(0, \frac{1}{2})$ $(2, 1\frac{1}{4})$ $(2, 2)$ $(3, \frac{1}{2})$

a. Explain how you know.

b. Give two other points that fall on this line.

3. Complete the tables for the given rules.

Line ℓ

Rule: *Halve x, and then add 1*

x	y	(x, y)
0		
1		
2		
3		

Line m

Rule: *Halve x, and then add $1\frac{1}{4}$*

x	y	(x, y)
0		
1		
2		
3		

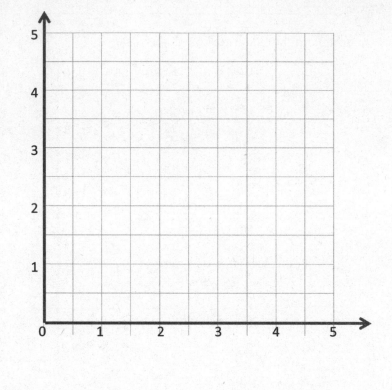

a. Draw each line on the coordinate plane above.

b. Compare and contrast these lines.

c. Based on the patterns you see, predict what the line for the rule *halve x, and then subtract 1* would look like. Draw your prediction on the plane above.

4. Circle the point(s) that the line for the rule *multiply x by $\frac{3}{4}$, and then subtract $\frac{1}{2}$* would contain.

$(1, \frac{1}{4})$ $\qquad$ $(2, \frac{1}{4})$ $\qquad$ $(3, 1\frac{3}{4})$ $\qquad$ $(3, 1)$

a. Explain how you know.

b. Give two other points that fall on this line.

Lesson 11: Analyze number patterns created from mixed operations.

©2015 Great Minds eureka-math.org
G5-M6-SE-BK3-1.3.1-02.2016

Line ℓ

Rule: *Triple x*

x	y	(x, y)
0		
1		
2		
4		

Line m

Rule: *Triple x, and then add 3*

x	y	(x, y)
0		
1		
2		
3		

Line n

Rule: *Triple x, and then subtract 2*

x	y	(x, y)
1		
2		
3		
4		

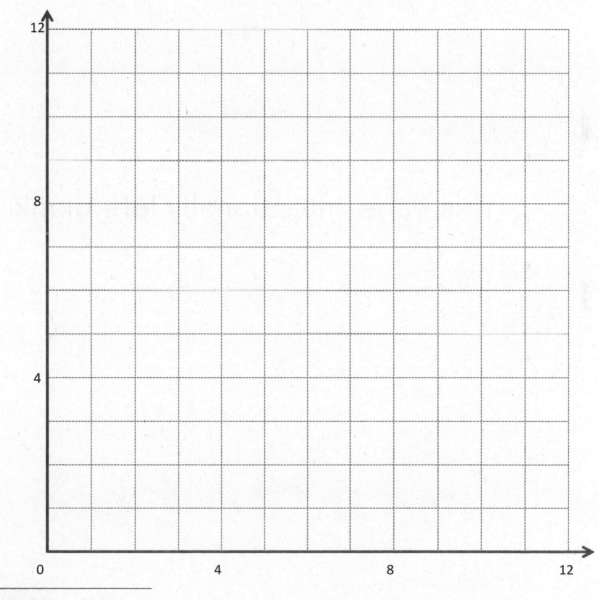

coordinate plane

This page intentionally left blank

Name _____   Date _____

1. Write a rule for the line that contains the points $(0, \frac{3}{4})$ and $(2\frac{1}{2}, 3\frac{1}{4})$.

a. Identify 2 more points on this line. Draw the line on the grid below.

Point	x	y	(x, y)
B			
C			

b. Write a rule for a line that is parallel to $\overleftrightarrow{BC}$ and goes through point $(1, \frac{1}{4})$.

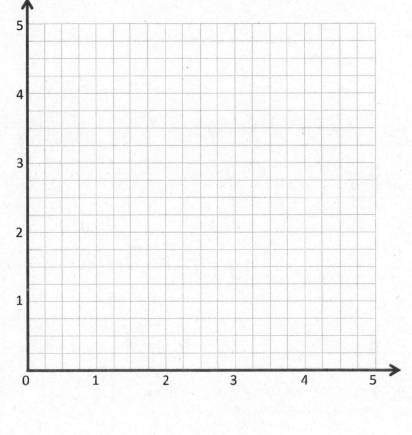

2. Create a rule for the line that contains the points $(1, \frac{1}{4})$ and $(3, \frac{3}{4})$.

a. Identify 2 more points on this line. Draw the line on the grid on the right.

Point	x	y	(x, y)
G			
H			

b. Write a rule for a line that passes through the origin and lies between $\overrightarrow{BC}$ and $\overrightarrow{GH}$.

3. Create a rule for a line that contains the point $(\frac{1}{4}, 1\frac{1}{4})$ using the operation or description below. Then, name 2 other points that would fall on each line.

a. Addition: _____

Point	x	y	(x, y)
T			
U			

b. A line parallel to the x-axis: _____

Point	x	y	(x, y)
G			
H			

c. Multiplication: _____

Point	x	y	(x, y)
A			
B			

d. A line parallel to the y-axis: _____

Point	x	y	(x, y)
V			
W			

e. Multiplication with addition: _____

Point	x	y	(x, y)
R			
S			

4. Mrs. Boyd asked her students to give a rule that could describe a line that contains the point (0.6, 1.8). Avi said the rule could be *multiply x by 3*. Ezra claims this could be a vertical line, and the rule could be *x is always 0.6*. Erik thinks the rule could be *add 1.2 to x*. Mrs. Boyd says that all the lines they are describing could describe a line that contains the point she gave. Explain how that is possible, and draw the lines on the coordinate plane to support your response.

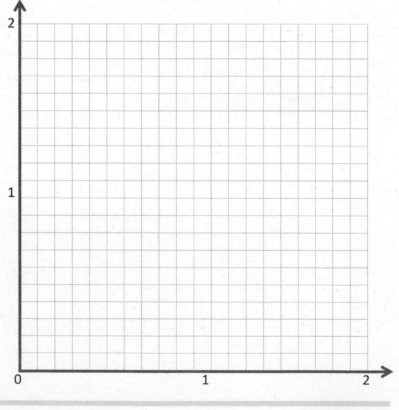

Lesson 12: Create a rule to generate a number pattern, and plot the points.

Extension:

5. Create a mixed operation rule for the line that contains the points (0, 1) and (1, 3).

a. Identify 2 more points, O and P, on this line. Draw the line on the grid.

Point	x	y	(x, y)
O			
P			

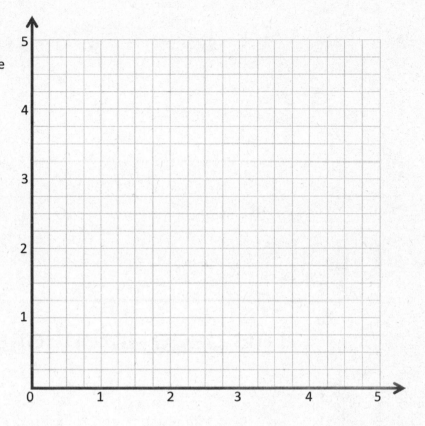

b. Write a rule for a line that is parallel to $\overleftrightarrow{OP}$ and goes through point $(1, 2\frac{1}{2})$.

Lesson 12:   Create a rule to generate a number pattern, and plot the points.

87

©2015 Great Minds eureka-math.org
G5-M6-SE-BK3-1.3.1-02.2016

This page intentionally left blank

Name _____   Date _____

1. Write a rule for the line that contains the points $(0, \frac{1}{4})$ and $(2\frac{1}{2}, 2\frac{3}{4})$.

   a. Identify 2 more points on this line. Draw the line on the grid below.

Point	x	y	(x, y)
B			
C			

   b. Write a rule for a line that is parallel to $\overleftrightarrow{BC}$ and goes through point $(1, 2\frac{1}{4})$.

   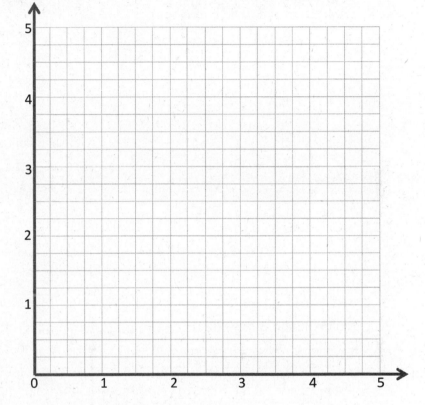

2. Give the rule for the line that contains the points $(1, 2\frac{1}{2})$ and $(2\frac{1}{2}, 2\frac{1}{2})$.

   a. Identify 2 more points on this line. Draw the line on the grid above.

Point	x	y	(x, y)
G			
H			

   b. Write a rule for a line that is parallel to $\overleftrightarrow{GH}$.

3. Give the rule for a line that contains the point $(\frac{3}{4}, 1\frac{1}{2})$ using the operation or description below. Then, name 2 other points that would fall on each line.

a. Addition: _____

Point	x	y	(x, y)
T			
U			

b. A line parallel to the x-axis: _____

Point	x	y	(x, y)
G			
H			

c. Multiplication: _____

Point	x	y	(x, y)
A			
B			

d. A line parallel to the y-axis: _____

Point	x	y	(x, y)
V			
W			

e. Multiplication with addition: _____

Point	x	y	(x, y)
R			
S			

4. On the grid, two lines intersect at (1.2, 1.2). If line a passes through the origin and line b contains the point (1.2, 0), write a rule for line a and line b.

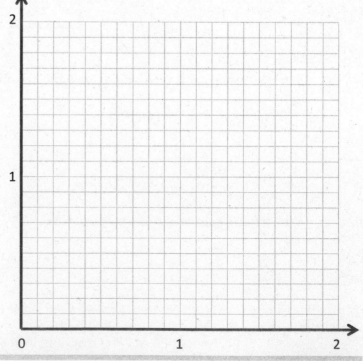

Lesson 12:   Create a rule to generate a number pattern, and plot the points.

EUREKA
MATH

Line l

Rule: _____

Line m

Rule: _____

Point	x	y	(x, y)
A	$1\frac{1}{2}$	3	$(1\frac{1}{2}, 3)$
B			
C			
D			

Point	x	y	(x, y)
A			
E			
F			
G			

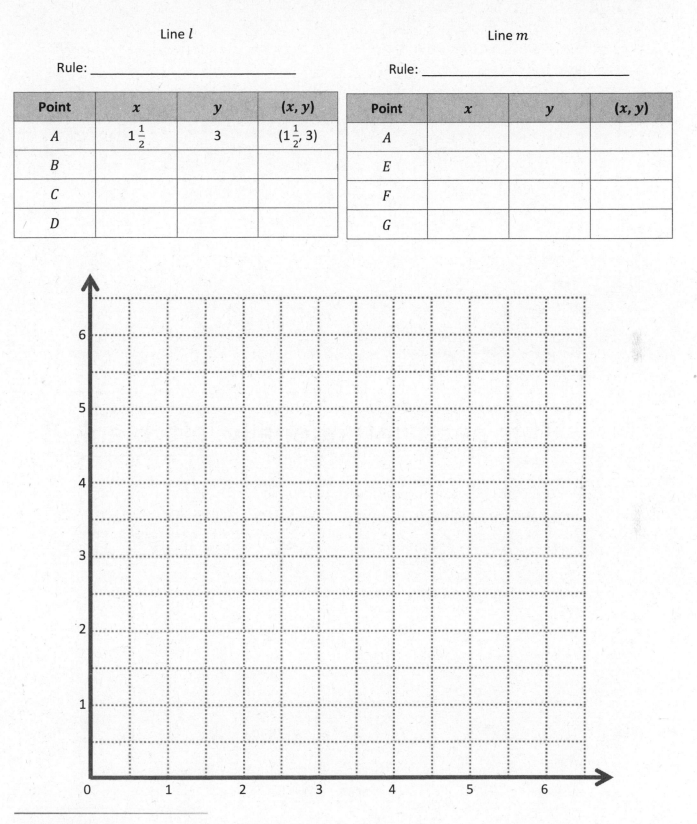

coordinate plane

Lesson 12:   Create a rule to generate a number pattern, and plot the points.

91

©2015 Great Minds eureka-math.org
G5-M6-SE-BK3-1.3.1-02.2016

This page intentionally left blank

Name _____   Date _____

1. Use a right angle template and straightedge to draw at least four sets of parallel lines in the space below.

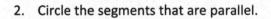

2. Circle the segments that are parallel.

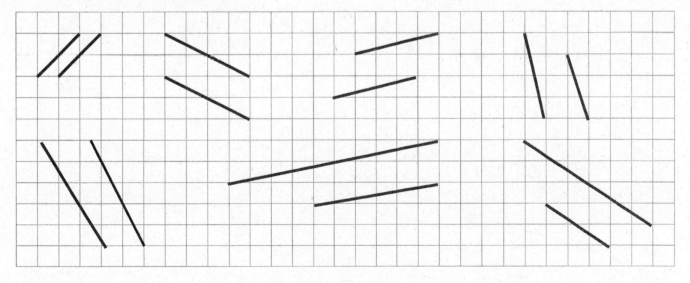

Lesson 13:   Construct parallel line segments on a rectangular grid.

93

©2015 Great Minds eureka-math.org
G5-M6-SE-BK3-1.3.1-02.2016

3. Use your straightedge to draw a segment parallel to each segment through the given point.

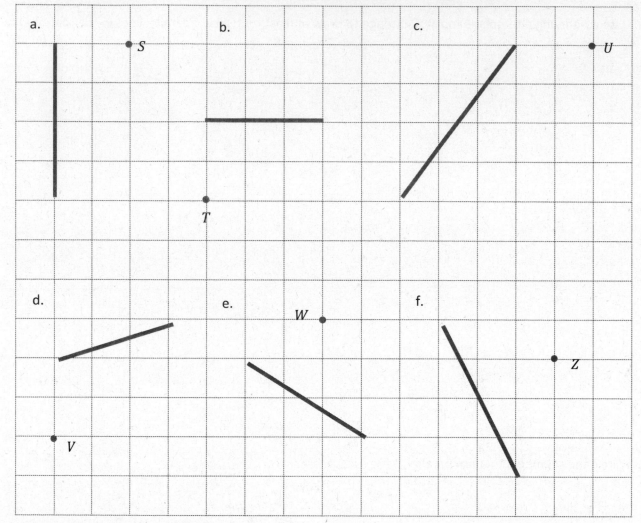

4. Draw 2 different lines parallel to line b.

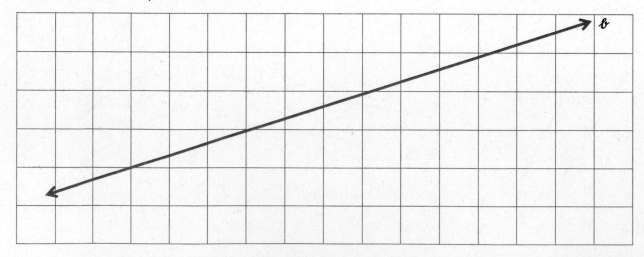

Lesson 13: Construct parallel line segments on a rectangular grid.

©2015 Great Minds eureka-math.org
G5-M6-SE-BK3-1.3.1-02.2016

Name _____   Date _____

1. Use your right angle template and straightedge to draw at least three sets of parallel lines in the space below.

2. Circle the segments that are parallel.

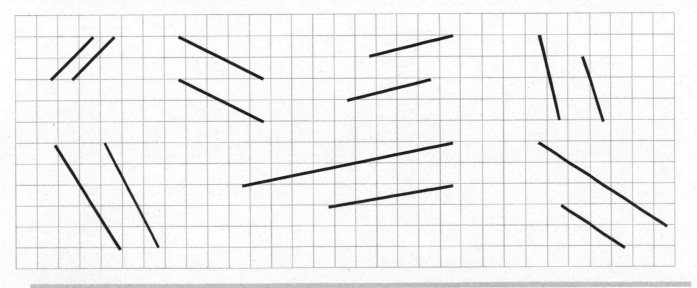

Lesson 13:   Construct parallel line segments on a rectangular grid.

©2015 Great Minds eureka-math.org
G5-M6-SE-BK3-1.3.1-02.2016

3. Use your straightedge to draw a segment parallel to each segment through the given point.

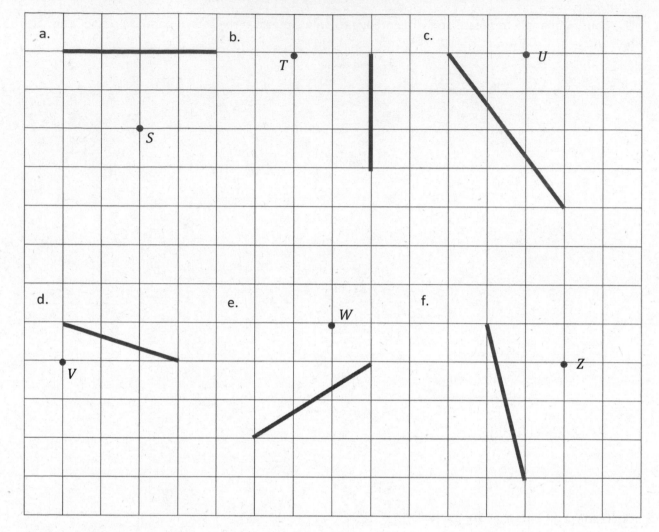

4. Draw 2 different lines parallel to line 𝑏.

Lesson 13:   Construct parallel line segments on a rectangular grid.

a.↓   b.↓   c.↓   d.↓

e.→

f.↓   g.→   h.→

rectangles

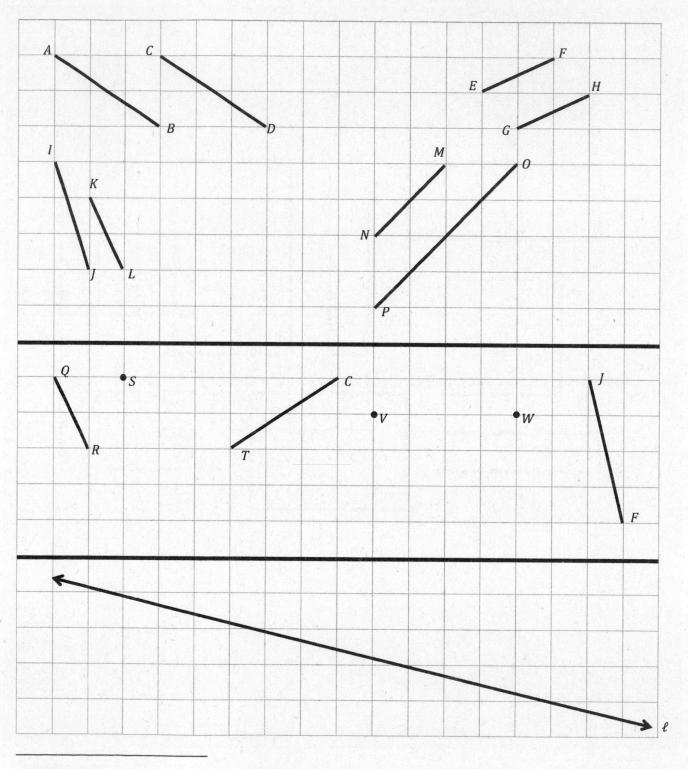

recording sheet

Lesson 13:   Construct parallel line segments on a rectangular grid.

©2015 Great Minds eureka-math.org
G5-M6-SE-BK3-1.3.1-02.2016

EUREKA
MATH™

Name _____   Date _____

1. Use the coordinate plane below to complete the following tasks.

a. Identify the locations of P and R. P: (_____, _____) R: (_____, _____)

b. Draw $\overleftrightarrow{PR}$.

c. Plot the following coordinate pairs on the plane.

S: (6, 7) T: (11, 9)

d. Draw $\overleftrightarrow{ST}$.

e. Circle the relationship between $\overleftrightarrow{PR}$ and $\overleftrightarrow{ST}$. $\overleftrightarrow{PR} \perp \overleftrightarrow{ST}$ $\overleftrightarrow{PR} \parallel \overleftrightarrow{ST}$

f. Give the coordinates of a pair of points, U and V, such that $\overleftrightarrow{UV} \parallel \overleftrightarrow{PR}$.

U: (_____, _____) V: (_____, _____)

g. Draw $\overleftrightarrow{UV}$.

©2015 Great Minds eureka-math.org
G5-M6-SE-BK3-1.3.1-02.2016

2. Use the coordinate plane below to complete the following tasks.

a. Identify the locations of E and F. E: (_____, _____) F: (_____, _____)

b. Draw $\overleftrightarrow{EF}$.

c. Generate coordinate pairs for L and M, such that $\overleftrightarrow{EF} \parallel \overleftrightarrow{LM}$.

L: (____, ____) M: (____, ____)

d. Draw $\overleftrightarrow{LM}$.

e. Explain the pattern you made use of when generating coordinate pairs for L and M.

f. Give the coordinates of a point, H, such that $\overleftrightarrow{EF} \parallel \overleftrightarrow{GH}$.

G: $(1\frac{1}{2}, 4)$ H: (____, ____)

g. Explain how you chose the coordinates for H.

Lesson 14:   Construct parallel line segments, and analyze relationships of the
                coordinate pairs.

©2015 Great Minds eureka-math.org
G5-M6-SE-BK3-1.3.1-02.2016

Name _____   Date _____

1. Use the coordinate plane below to complete the following tasks.

a. Identify the locations of M and N. M: (_____, _____) N: (_____, _____)

b. Draw $\overleftrightarrow{MN}$.

c. Plot the following coordinate pairs on the plane.

J: (5, 7) K: (8, 5)

d. Draw $\overrightarrow{JK}$.

e. Circle the relationship between $\overleftrightarrow{MN}$ and $\overrightarrow{JK}$. $\overleftrightarrow{MN} \perp \overrightarrow{JK}$ $\overleftrightarrow{MN} \parallel \overrightarrow{JK}$

f. Give the coordinates of a pair of points, F and G, such that $\overleftrightarrow{FG} \parallel \overleftrightarrow{MN}$.

F: (_____, _____) G: (_____, _____)

g. Draw $\overleftrightarrow{FG}$.

2. Use the coordinate plane below to complete the following tasks.

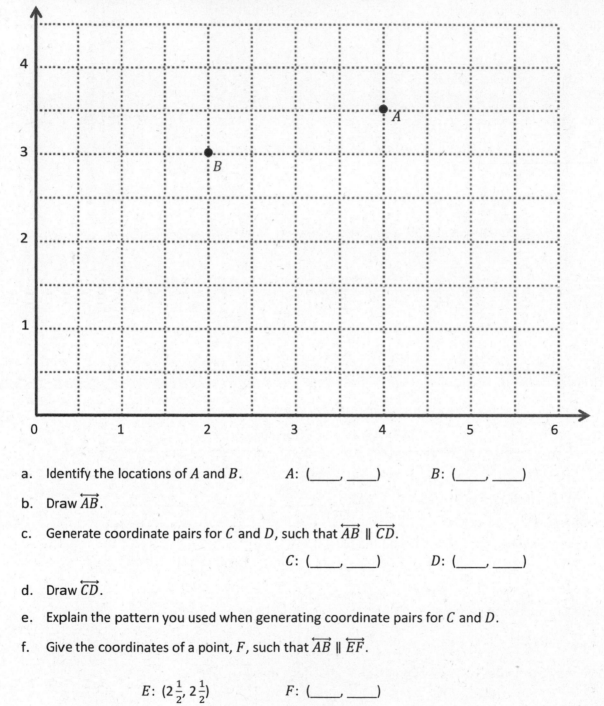

a. Identify the locations of A and B. A: (___, ___) B: (___, ___)

b. Draw $\overleftrightarrow{AB}$.

c. Generate coordinate pairs for C and D, such that $\overleftrightarrow{AB} \parallel \overleftrightarrow{CD}$.

                                    C: (___, ___) D: (___, ___)

d. Draw $\overleftrightarrow{CD}$.

e. Explain the pattern you used when generating coordinate pairs for C and D.

f. Give the coordinates of a point, F, such that $\overleftrightarrow{AB} \parallel \overleftrightarrow{EF}$.

        E: $(2\frac{1}{2}, 2\frac{1}{2})$ F: (___, ___)

g. Explain how you chose the coordinates for F.

Lesson 14:   Construct parallel line segments, and analyze relationships of the coordinate pairs.

©2015 Great Minds eureka-math.org
G5-M6-SE-BK3-1.3.1-02.2016

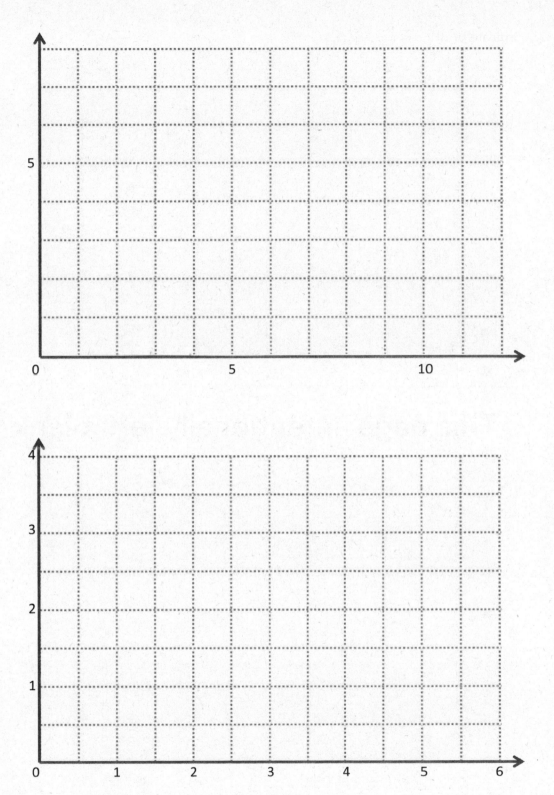

coordinate plane

Lesson 14: Construct parallel line segments, and analyze relationships of the coordinate pairs.

103

©2015 Great Minds eureka-math.org
G5-M6-SE-BK3-1.3.1-02.2016

This page intentionally left blank

Name _____ Date _____

1. Circle the pairs of segments that are perpendicular.

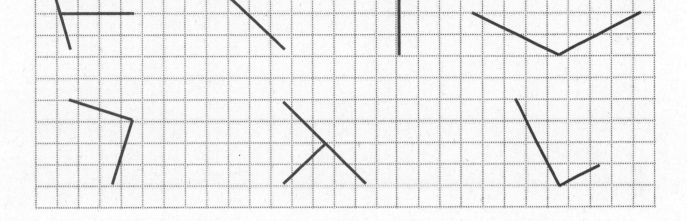

2. In the space below, use your right triangle templates to draw at least 3 different sets of perpendicular lines.

EUREKA MATH

Lesson 15:   Construct perpendicular line segments on a rectangular grid.

105

©2015 Great Minds eureka-math.org
G5-M6-SE-BK3-1.3.1-02.2016

3. Draw a segment perpendicular to each given segment. Show your thinking by sketching triangles as needed.

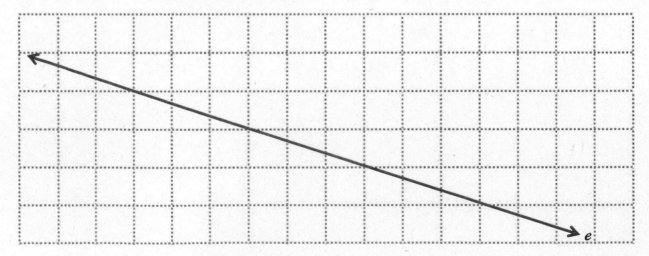

a.

b.

c.

d.

4. Draw 2 different lines perpendicular to line *e*.

e

Lesson 15:   Construct perpendicular line segments on a rectangular grid.

Name _____ Date _____

1. Circle the pairs of segments that are perpendicular.

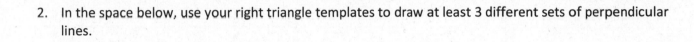

2. In the space below, use your right triangle templates to draw at least 3 different sets of perpendicular lines.

Lesson 15:   Construct perpendicular line segments on a rectangular grid.

©2015 Great Minds eureka-math.org
G5-M6-SE-BK3-1.3.1-02.2016

107

3. Draw a segment perpendicular to each given segment. Show your thinking by sketching triangles as needed.

a.

b.

c.

d.

4. Draw 2 different lines perpendicular to line *b*.

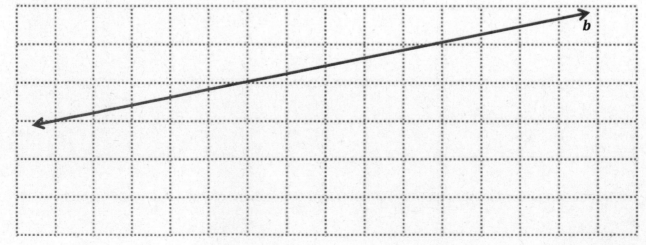

         Lesson 15:   Construct perpendicular line segments on a rectangular grid.

©2015 Great Minds eureka-math.org
G5-M6-SE-BK3-1.3.1-02.2016

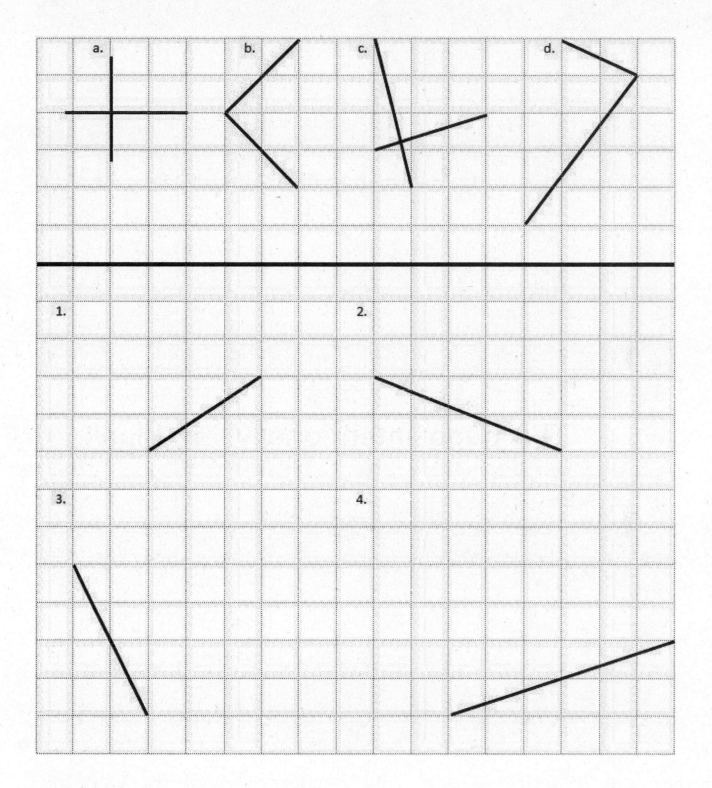

recording sheet

EUREKA MATH

Lesson 15:   Construct perpendicular line segments on a rectangular grid.

109

©2015 Great Minds eureka-math.org
G5-M6-SE-BK3-1.3.1-02.2016

This page intentionally left blank

Name _____   Date _____

1. Use the coordinate plane below to complete the following tasks.

    a. Draw $\overline{AB}$.

    b. Plot point C (0, 8).

    c. Draw $\overline{AC}$.

    d. Explain how you know $\angle CAB$ is a right angle without measuring it.

    e. Sean drew the picture below to find a segment perpendicular to $\overline{AB}$. Explain why Sean is correct.

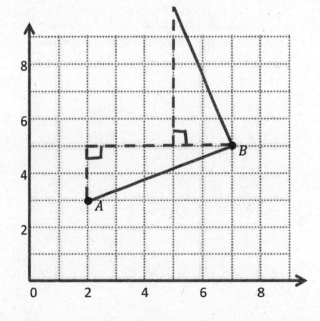

Lesson 16:   Construct perpendicular line segments, and analyze relationships of the coordinate pairs.

111

2. Use the coordinate plane below to complete the following tasks.

    a. Draw $\overline{QT}$.

    b. Plot point R $(2, 6\frac{1}{2})$.

    c. Draw $\overline{QR}$.

    d. Explain how you know $\angle RQT$ is a right angle without measuring it.

    e. Compare the coordinates of points Q and T. What is the difference of the x-coordinates? The y-coordinates?

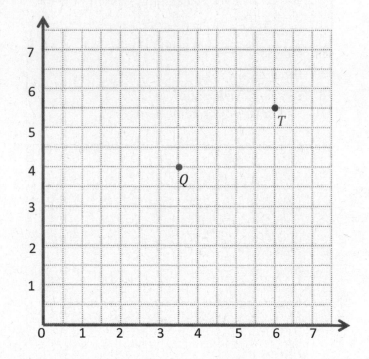

    f. Compare the coordinates of points Q and R. What is the difference of the x-coordinates? The y-coordinates?

    g. What is the relationship of the differences you found in parts (e) and (f) to the triangles of which these two segments are a part?

3. $\overleftrightarrow{EF}$ contains the following points. E: (4, 1) F: (8, 7)

   Give the coordinates of a pair of points G and H, such that $\overleftrightarrow{EF} \perp \overleftrightarrow{GH}$.

   G: (_____, _____) H: (_____, _____)

Lesson 16:   Construct perpendicular line segments, and analyze relationships of the coordinate pairs.

©2015 Great Minds eureka-math.org
G5-M6-SE-BK3-1.3.1-02.2016

Name _____   Date _____

1. Use the coordinate plane below to complete the following tasks.

    a. Draw $\overline{PQ}$.

    b. Plot point R (3, 8).

    c. Draw $\overline{PR}$.

    d. Explain how you know $\angle RPQ$ is a right angle without measuring it.

    e. Compare the coordinates of points P and Q. What is the difference of the x-coordinates? The y-coordinates?

    f. Compare the coordinates of points P and R. What is the difference of the x-coordinates? The y-coordinates?

    g. What is the relationship of the differences you found in parts (e) and (f) to the triangles of which these two segments are a part?

Lesson 16:   Construct perpendicular line segments, and analyze relationships of the coordinate pairs.

©2015 Great Minds eureka-math.org
G5-M6-SE-BK3-1.3.1-02.2016

113

2. Use the coordinate plane below to complete the following tasks.

    a. Draw $\overline{CB}$.

    b. Plot point $D\left(\frac{1}{2}, 5\frac{1}{2}\right)$.

    c. Draw $\overline{CD}$.

    d. Explain how you know $\angle DCB$ is a right angle without measuring it.

    e. Compare the coordinates of points C and B. What is the difference of the x-coordinates? The y-coordinates?

    f. Compare the coordinates of points C and D. What is the difference of the x-coordinates? The y-coordinates?

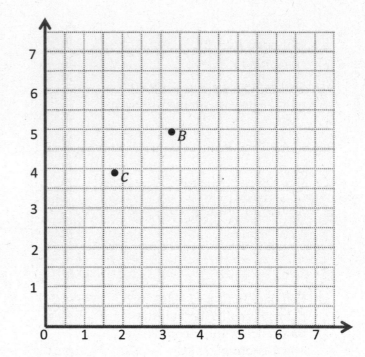

    g. What is the relationship of the differences you found in parts (e) and (f) to the triangles of which these two segments are a part?

3. $\overleftrightarrow{ST}$ contains the following points. S: (2, 3) T: (9, 6)

    Give the coordinates of a pair of points, U and V, such that $\overleftrightarrow{ST} \perp \overleftrightarrow{UV}$.

    U: (_____, _____) V: (_____, _____)

Lesson 16:   Construct perpendicular line segments, and analyze relationships of the coordinate pairs.

©2015 Great Minds eureka-math.org
G5-M6-SE-BK3-1.3.1-02.2016

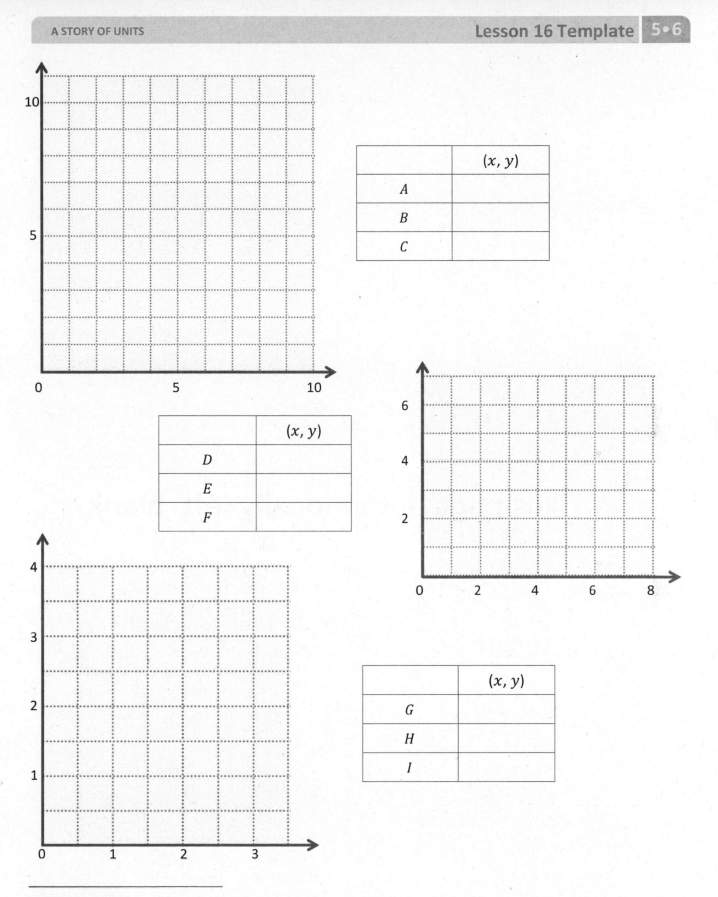

coordinate plane

Lesson 16:   Construct perpendicular line segments, and analyze relationships of
the coordinate pairs.

©2015 Great Minds eureka-math.org
G5-M6-SE-BK3-1.3.1-02.2016

115

This page intentionally left blank

Name _____   Date _____

1. Draw to create a figure that is symmetric about $\overleftrightarrow{AD}$.

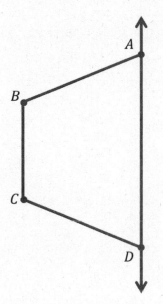

2. Draw precisely to create a figure that is symmetric about $\overleftrightarrow{HI}$.

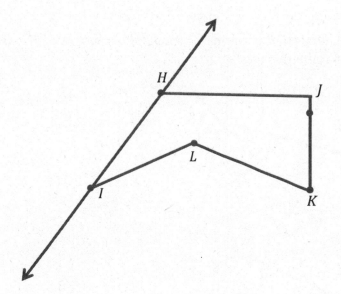

3. Complete the following construction in the space below.

   a. Plot 3 non-collinear points, D, E, and F.

   b. Draw $\overline{DE}$, $\overline{EF}$, and $\overleftrightarrow{DF}$.

   c. Plot point G, and draw the remaining sides, such that quadrilateral $DEFG$ is symmetric about $\overleftrightarrow{DF}$.

4. Stu says that quadrilateral $HIJK$ is symmetric about $\overrightarrow{HJ}$ because $IL = LK$. Use your tools to determine Stu's mistake. Explain your thinking.

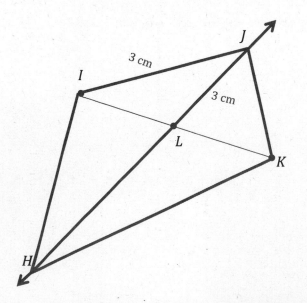

Lesson 17:   Draw symmetric figures using distance and angle measure from the line of symmetry.

©2015 Great Minds eureka-math.org
G5-M6-SE-BK3-1.3.1-02.2016

Name _____ Date _____

1. Draw to create a figure that is symmetric about $\overleftrightarrow{DE}$.

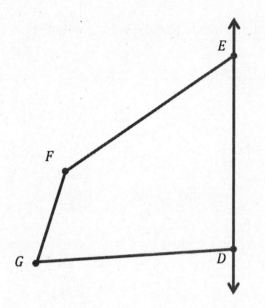

2. Draw to create a figure that is symmetric about $\overleftrightarrow{LM}$.

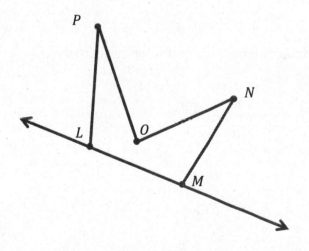

Lesson 17:   Draw symmetric figures using distance and angle measure from the line of symmetry.

©2015 Great Minds eureka-math.org
G5-M6-SE-BK3-1.3.1-02.2016

119

3. Complete the following construction in the space below.

   a. Plot 3 non-collinear points, G, H, and I.

   b. Draw $\overline{GH}$, $\overline{HI}$, and $\overleftrightarrow{IG}$.

   c. Plot point J, and draw the remaining sides, such that quadrilateral $GHIJ$ is symmetric about $\overleftrightarrow{IG}$.

4. In the space below, use your tools to draw a symmetric figure about a line.

Lesson 17: Draw symmetric figures using distance and angle measure from the line of symmetry.

©2015 Great Minds eureka-math.org
G5-M6-SE-BK3-1.3.1-02.2016

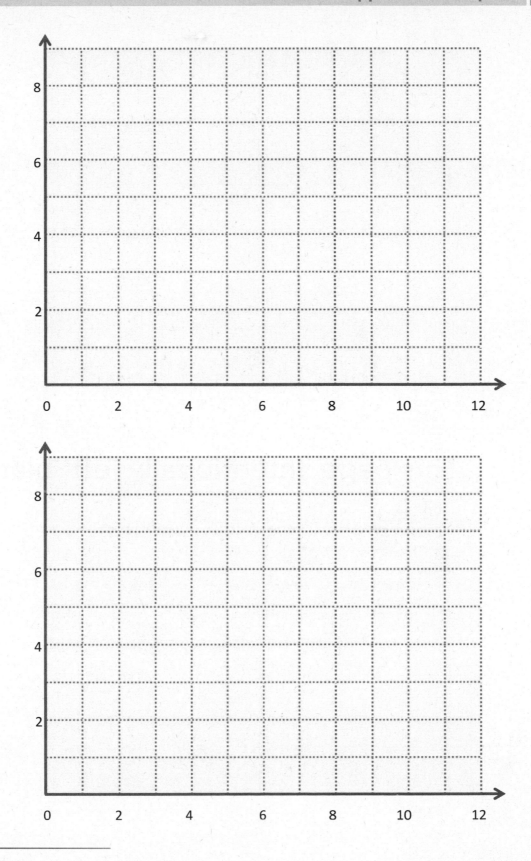

coordinate plane

Lesson 17: Draw symmetric figures using distance and angle measure from the line of symmetry.

©2015 Great Minds eureka-math.org
G5-M6-SE-BK3-1.3.1-02.2016

121

This page intentionally left blank

Name _____   Date _____

1. Use the plane to the right to complete the following tasks.

    a. Draw a line t whose rule is *y is always 0.7*.

    b. Plot the points from Table A on the grid in order. Then, draw line segments to connect the points.

Table A

(x, y)
(0.1, 0.5)
(0.2, 0.3)
(0.3, 0.5)
(0.5, 0.1)
(0.6, 0.2)
(0.8, 0.2)
(0.9, 0.1)
(1.1, 0.5)
(1.2, 0.3)
(1.3, 0.5)

Table B

(x, y)

    c. Complete the drawing to create a figure that is symmetric about line t. For each point in Table A, record the corresponding point on the other side of the line of symmetry in Table B.

    d. Compare the y-coordinates in Table A with those in Table B. What do you notice?

    e. Compare the x-coordinates in Table A with those in Table B. What do you notice?

2. This figure has a second line of symmetry. Draw the line on the plane, and write the rule for this line.

©2015 Great Minds eureka-math.org
G5-M6-SE-BK3-1.3.1-02.2016

3. Use the plane below to complete the following tasks.

   a. Draw a line u whose rule is *y is equal to* $x + \frac{1}{4}$.

   b. Construct a figure with a total of 6 points, all on the same side of the line.

   c. Record the coordinates of each point, in the order in which they were drawn, in Table A.

   d. Swap your paper with a neighbor, and have her complete parts (e–f), below.

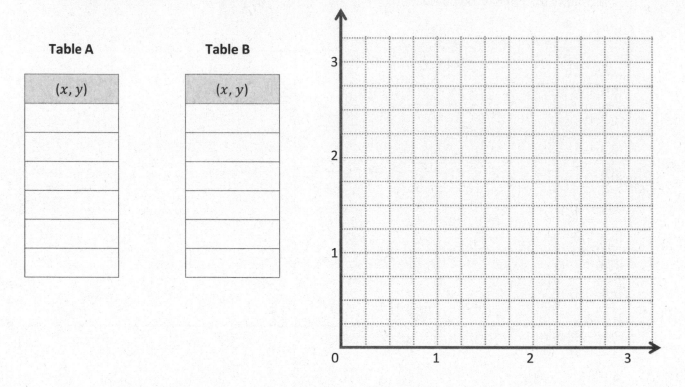

   e. Complete the drawing to create a figure that is symmetric about u. For each point in Table A, record the corresponding point on the other side of the line of symmetry in Table B.

   f. Explain how you found the points symmetric to your partner's about u.

Lesson 18:   Draw symmetric figures on the coordinate plane.

©2015 Great Minds eureka-math.org
G5-M6-SE-BK3-1.3.1-02.2016

Name _____   Date _____

1. Use the plane to the right to complete the following tasks.

   a. Draw a line *s* whose rule is *x is always 5*.

   b. Plot the points from Table A on the grid in order. Then, draw line segments to connect the points in order.

Table A		Table B
(x, y)		(x, y)
(1, 13)		
(1, 12)		
(2, 10)		
(4, 9)		
(4, 3)		
(1, 2)		
(5, 2)		

   c. Complete the drawing to create a figure that is symmetric about line *s*. For each point in Table A, record the symmetric point on the other side of *s*.

   d. Compare the *y*-coordinates in Table A with those in Table B. What do you notice?

   e. Compare the *x*-coordinates in Table A with those in Table B. What do you notice?

2. Use the plane to the right to complete the following tasks.

   a. Draw a line p whose rule is, *y is equal to x*.

   b. Plot the points from Table A on the grid in order. Then, draw line segments to connect the points.

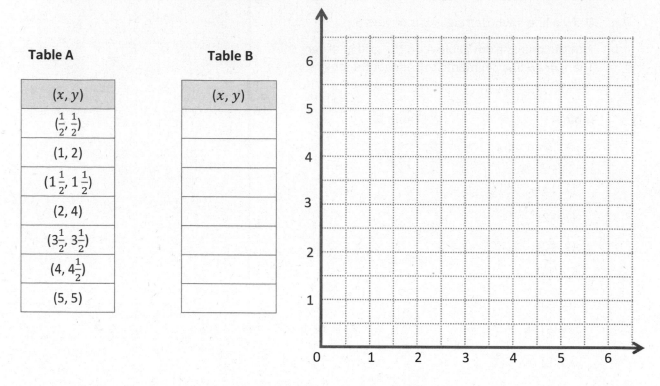

Table A

(x, y)
$(\frac{1}{2}, \frac{1}{2})$
$(1, 2)$
$(1\frac{1}{2}, 1\frac{1}{2})$
$(2, 4)$
$(3\frac{1}{2}, 3\frac{1}{2})$
$(4, 4\frac{1}{2})$
$(5, 5)$

Table B

(x, y)

   c. Complete the drawing to create a figure that is symmetric about line p. For each point in Table A, record the symmetric point on the other side of the line p in Table B.

   d. Compare the y-coordinates in Table A with those in Table B. What do you notice?

   e. Compare the x-coordinates in Table A with those in Table B. What do you notice?

Lesson 18:   Draw symmetric figures on the coordinate plane.

©2015 Great Minds eureka-math.org
G5-M6-SE-BK3-1.3.1-02.2016

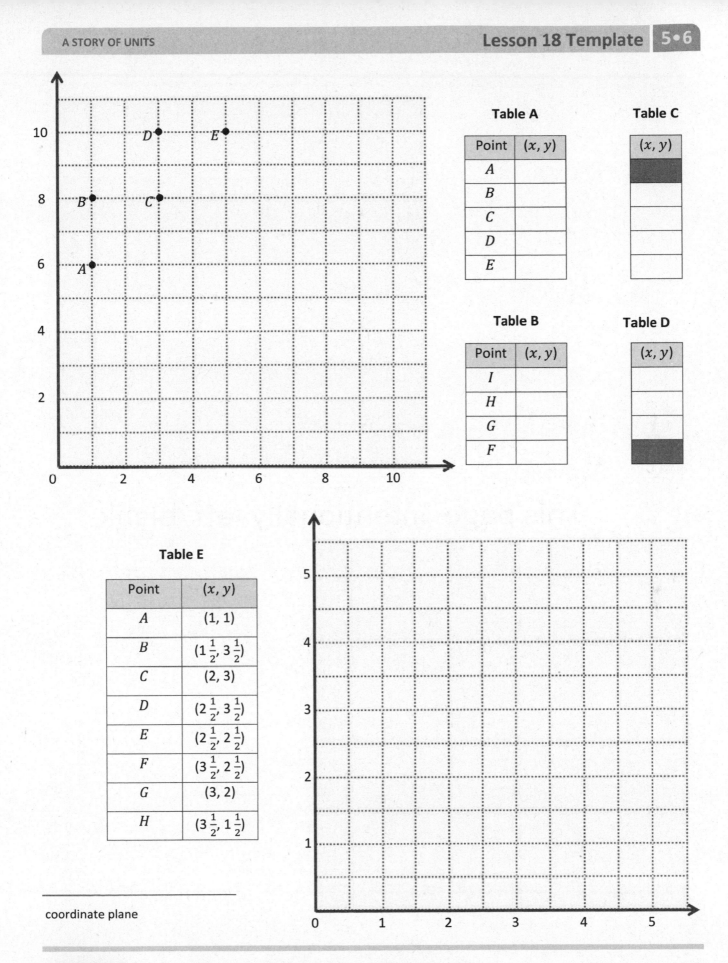

Table A

Point	(x, y)
A	
B	
C	
D	
E	

Table C

(x, y)

Table B

Point	(x, y)
I	
H	
G	
F	

Table D

(x, y)

Table E

Point	(x, y)
A	$(1, 1)$
B	$(1\frac{1}{2}, 3\frac{1}{2})$
C	$(2, 3)$
D	$(2\frac{1}{2}, 3\frac{1}{2})$
E	$(2\frac{1}{2}, 2\frac{1}{2})$
F	$(3\frac{1}{2}, 2\frac{1}{2})$
G	$(3, 2)$
H	$(3\frac{1}{2}, 1\frac{1}{2})$

coordinate plane

This page intentionally left blank

Name _____ Date _____

1. The line graph below tracks the rain accumulation, measured every half hour, during a rainstorm that began at 2:00 p.m. and ended at 7:00 p.m. Use the information in the graph to answer the questions that follow.

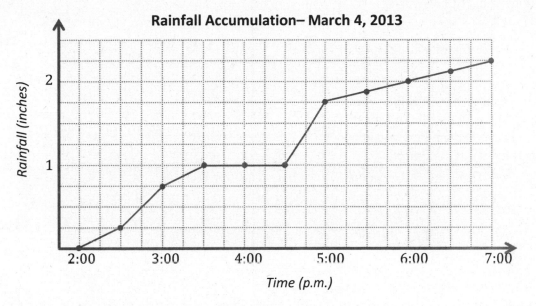

Rainfall Accumulation– March 4, 2013

Rainfall (inches)

Time (p.m.)

a. How many inches of rain fell during this five-hour period?

b. During which half-hour period did $\frac{1}{2}$ inch of rain fall? Explain how you know.

c. During which half-hour period did rain fall most rapidly? Explain how you know.

d. Why do you think the line is horizontal between 3:30 p.m. and 4:30 p.m.?

e. For every inch of rain that fell here, a nearby community in the mountains received a foot and a half of snow. How many inches of snow fell in the mountain community between 5:00 p.m. and 7:00 p.m.?

2. Mr. Boyd checks the gauge on his home's fuel tank on the first day of every month. The line graph to the right was created using the data he collected.

    a. According to the graph, during which month(s) does the amount of fuel decrease most rapidly?

    b. The Boyds took a month-long vacation. During which month did this most likely occur? Explain how you know using the data in the graph.

    c. Mr. Boyd's fuel company filled his tank once this year. During which month did this most likely occur? Explain how you know.

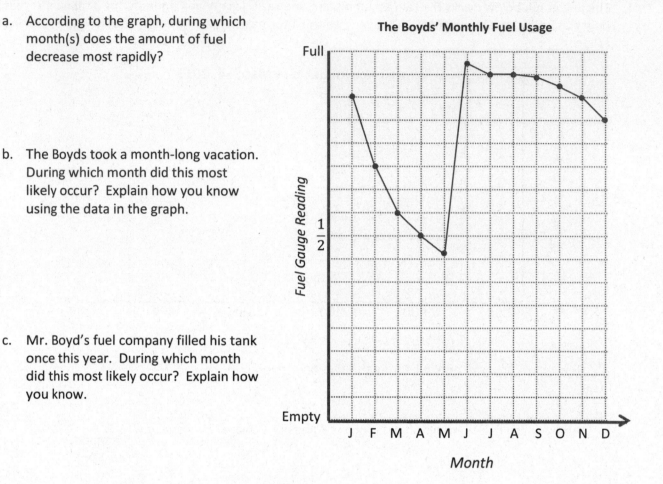

    d. The Boyd family's fuel tank holds 284 gallons of fuel when full. How many gallons of fuel did the Boyds use in February?

    e. Mr. Boyd pays $3.54 per gallon of fuel. What is the cost of the fuel used in February and March?

Lesson 19: Plot data on line graphs and analyze trends.

EUREKA MATH

Name _____ Date _____

1. The line graph below tracks the balance of Howard's checking account, at the end of each day, between May 12 and May 26. Use the information in the graph to answer the questions that follow.

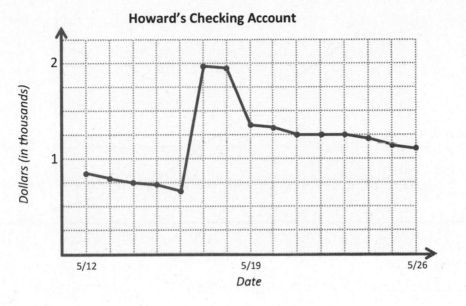

Howard's Checking Account

a. About how much money does Howard have in his checking account on May 21?

b. If Howard spends $250 from his checking account on May 26, about how much money will he have left in his account?

c. Explain what happened with Howard's money between May 21 and May 23.

d. Howard received a payment from his job that went directly into his checking account. On which day did this most likely occur? Explain how you know.

e. Howard bought a new television during the time shown in the graph. On which day did this most likely occur? Explain how you know.

©2015 Great Minds eureka-math.org
G5-M6-SE-BK3-1.3.1-02.2016

2. The line graph below tracks Santino's time at the beginning and end of each part of a triathlon. Use the information in the graph to answer the questions that follow.

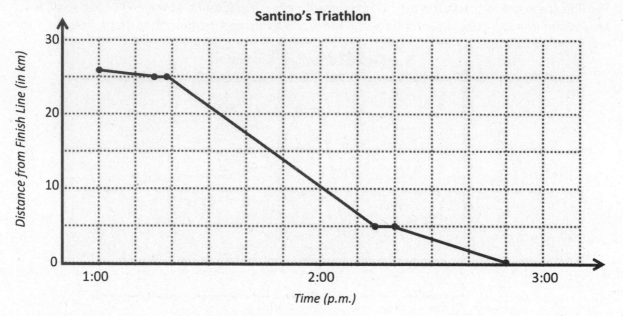

Santino's Triathlon

a. How long does it take Santino to finish the triathlon?

b. To complete the triathlon, Santino first swims across a lake, then bikes through the city, and finishes by running around the lake. According to the graph, what was the distance of the running portion of the race?

c. During the race, Santino pauses to put on his biking shoes and helmet and then later to change into his running shoes. At what times did this most likely occur? Explain how you know.

d. Which part of the race does Santino finish most quickly? How do you know?

e. During which part of the triathlon is Santino racing most quickly? Explain how you know.

Lesson 19:   Plot data on line graphs and analyze trends.

©2015 Great Minds eureka-math.org
G5-M6-SE-BK3-1.3.1-02.2016

Name _____   Date _____

1. The line graph below tracks the total tomato production for one tomato plant. The total tomato production is plotted at the end of each of 8 weeks. Use the information in the graph to answer the questions that follow.

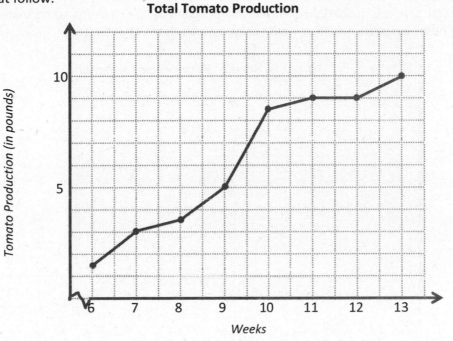

Total Tomato Production

a. How many pounds of tomatoes did this plant produce at the end of 13 weeks?

b. How many pounds of tomatoes did this plant produce from Week 7 to Week 11? Explain how you know.

c. Which one-week period showed the greatest change in tomato production? The least? Explain how you know.

d. During Weeks 6–8, Jason fed the tomato plant just water. During Weeks 8–10, he used a mixture of water and Fertilizer A, and in Weeks 10–13, he used water and Fertilizer B on the tomato plant. Compare the tomato production for these periods of time.

Lesson 20:   Use coordinate systems to solve real-world problems.

133

©2015 Great Minds eureka-math.org
G5-M6-SE-BK3-1.3.1-02.2016

2. Use the story context below to sketch a line graph. Then, answer the questions that follow.

The number of fifth-grade students attending Magnolia School has changed over time. The school opened in 2006 with 156 students in the fifth grade. The student population grew the same amount each year before reaching its largest class of 210 students in 2008. The following year, Magnolia lost one-seventh of its fifth graders. In 2010, the enrollment dropped to 154 students and remained constant in 2011. For the next two years, the enrollment grew by 7 students each year.

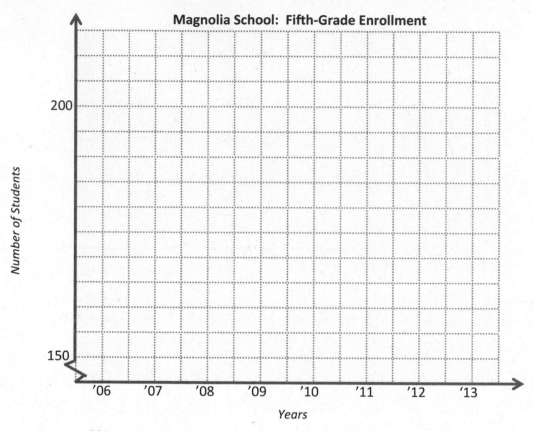

a. How many more fifth-grade students attended Magnolia in 2009 than in 2013?

b. Between which two consecutive years was there the greatest change in student population?

c. If the fifth-grade population continues to grow in the same pattern as in 2012 and 2013, in what year will the number of students match 2008's enrollment?

134

Lesson 20:   Use coordinate systems to solve real-world problems.

Name _____ Date _____

Use the graph to answer the questions.

Johnny left his home at 6 a.m. and kept track of the number of kilometers he traveled at the end of each hour of his trip. He recorded the data in a line graph.

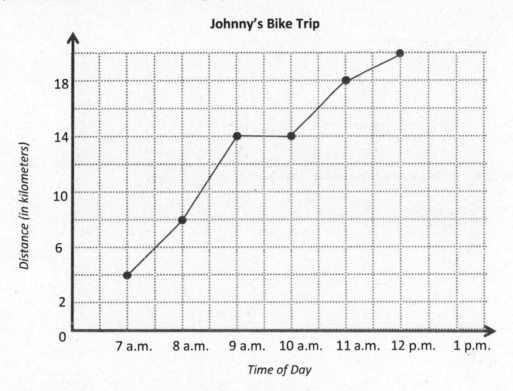

a. How far did Johnny travel in all? How long did it take?

b. Johnny took a one-hour break to have a snack and take some pictures. What time did he stop? How do you know?

Lesson 20:   Use coordinate systems to solve real-world problems.

135

©2015 Great Minds eureka-math.org
G5-M6-SE-BK3-1.3.1-02.2016

c. Did Johnny cover more distance before his break or after? Explain.

d. Between which two hours did Johnny ride 4 kilometers?

e. During which hour did Johnny ride the fastest? Explain how you know.

Lesson 20:   Use coordinate systems to solve real-world problems.

©2015 Great Minds eureka-math.org
G5-M6-SE-BK3-1.3.1-02.2016

EUREKA
MATH

Student _____ Team _____ Date _____ Problem 1

Pierre's Paper

Pierre folded a square piece of paper vertically to make two rectangles. Each rectangle had a perimeter of 39 inches. How long is each side of the original square? What is the area of the original square? What is the area of one of the rectangles?

Student _____ Team _____ Date _____ Problem 2

Shopping with Elise

Elise saved $184. She bought a scarf, a necklace, and a notebook. After her purchases, she still had $39.50. The scarf cost three-fifths the cost of the necklace, and the notebook was one-sixth as much as the scarf. What was the cost of each item? How much more did the necklace cost than the notebook?

Lesson 21:   Make sense of complex, multi-step problems, and persevere in solving them. Share and critique peer solutions.

137

©2015 Great Minds eureka-math.org
G5-M6-SE-BK3-1.3.1-02.2016

This page intentionally left blank

Student _____ Team _____ Date _____ Problem 3

The Hewitt's Carpet

The Hewitt family is buying carpet for two rooms. The dining room is a square that measures 12 feet on each side. The den is 9 yards by 5 yards. Mrs. Hewitt has budgeted $2,650 for carpeting both rooms. The green carpet she is considering costs $42.75 per square yard, and the brown carpet's price is $4.95 per square foot. What are the ways she can carpet the rooms and stay within her budget?

Student _____ Team _____ Date _____ Problem 4

AAA Taxi

AAA Taxi charges $1.75 for the first mile and $1.05 for each additional mile. How far could Mrs. Leslie travel for $20 if she tips the cab driver $2.50?

Lesson 21: Make sense of complex, multi-step problems, and persevere in solving them. Share and critique peer solutions.

©2015 Great Minds eureka-math.org
G5-M6-SE-BK3-1.3.1-02.2016

139

This page intentionally left blank

Student _____ Team _____ Date _____ Problem 5

Pumpkins and Squash

Three pumpkins and two squash weigh 27.5 pounds. Four pumpkins and three squash weigh 37.5 pounds. Each pumpkin weighs the same as the other pumpkins, and each squash weighs the same as the other squash. How much does each pumpkin weigh? How much does each squash weigh?

Student _____ Team _____ Date _____ Problem 6

Toy Cars and Trucks

Henry had 20 convertibles and 5 trucks in his miniature car collection. After Henry's aunt bought him some more miniature trucks, Henry found that one-fifth of his collection consisted of convertibles. How many trucks did his aunt buy?

EUREKA
MATH™

Lesson 21:   Make sense of complex, multi-step problems, and persevere in solving
them. Share and critique peer solutions.   141

©2015 Great Minds eureka-math.org
G5-M6-SE-BK3-1.3.1-02.2016

This page intentionally left blank

Student _____ Team _____ Date _____ Problem 7

Pairs of Scouts

Some girls in a Girl Scout troop are pairing up with some boys in a Boy Scout troop to practice square dancing. Two-thirds of the girls are paired with three-fifths of the boys. What fraction of the scouts are square dancing?

(Each pair is one Girl Scout and one Boy Scout. The pairs are only from these two troops.)

Student _____ Team _____ Date _____ Problem 8

Sandra's Measuring Cups

Sandra is making cookies that require $5\frac{1}{2}$ cups of oatmeal. She has only two measuring cups: a one-half cup and a three-fourths cup. What is the smallest number of scoops that she could make in order to get $5\frac{1}{2}$ cups?

Lesson 21:   Make sense of complex, multi-step problems, and persevere in solving them. Share and critique peer solutions.

©2015 Great Minds eureka-math.org
G5-M6-SE-BK3-1.3.1-02.2016

143

This page intentionally left blank

Student _____ Team _____ Date _____ Problem 9

Blue Squares

The dimensions of each successive blue square pictured to the right are half that of the previous blue square. The lower left blue square measures 6 inches by 6 inches.

    a.   Find the area of the shaded part.

    b.   Find the total area of the shaded and unshaded parts.

    c.   What fraction of the figure is shaded?

Lesson 21:   Make sense of complex, multi-step problems, and persevere in solving them. Share and critique peer solutions.

©2015 Great Minds eureka-math.org
G5-M6-SE-BK3-1.3.1-02.2016

145

This page intentionally left blank

Name _____   Date _____

1. Sara travels twice as far as Eli when going to camp. Ashley travels as far as Sara and Eli together. Hazel travels 3 times as far as Sara. In total, all four travel 888 miles to camp. How far does each of them travel?

Lesson 21:   Make sense of complex, multi-step problems, and persevere in solving them. Share and critique peer solutions.

©2015 Great Minds eureka-math.org
G5-M6-SE-BK3-1.3.1-02.2016

147

The following problem is a brainteaser for your enjoyment. It is intended to encourage working together and family problem-solving fun. It is not a required element of this homework assignment.

2. A man wants to take a goat, a bag of cabbage, and a wolf over to an island. His boat will only hold him and one animal or item. If the goat is left with the cabbage, he'll eat it. If the wolf is left with the goat, he'll eat it. How can the man transport all three to the island without anything being eaten?

Lesson 21: Make sense of complex, multi-step problems, and persevere in solving them. Share and critique peer solutions.

©2015 Great Minds eureka-math.org
G5-M6-SE-BK3-1.3.1-02.2016

Name _____   Date _____

Solve using any method. Show all your thinking.

1. Study this diagram showing all the squares. Fill in the table.

Figure	Area in Square Feet
1	1 ft^2
2	
3	
4	9 ft^2
5	
6	1 ft^2
7	
8	
9	
10	
11	

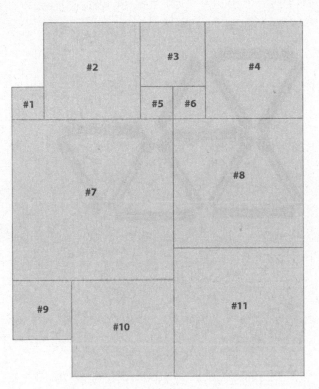

Lesson 22: Make sense of complex, multi-step problems, and persevere in solving them. Share and critique peer solutions.

©2015 Great Minds eureka-math.org
G5-M6-SE-BK3-1.3.1-02.2016

149

The following problem is a brainteaser for your enjoyment. It is intended to encourage working together and family problem-solving fun. It is not a required element of this homework assignment.

2.   Remove 3 matches to leave 3 triangles.

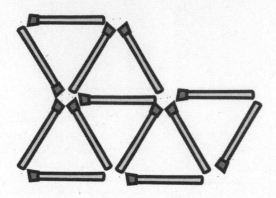

Lesson 22:   Make sense of complex, multi-step problems, and persevere in solving
                                        them. Share and critique peer solutions.

©2015 Great Minds eureka-math.org
G5-M6-SE-BK3-1.3.1-02.2016

EUREKA
MATH

Name _____ Date _____

1. In the diagram, the length of Figure S is $\frac{2}{3}$ the length of Figure T. If S has an area of 368 cm², find the perimeter of the figure.

S	T	

├─16 cm

Lesson 23:   Make sense of complex, multi-step problems, and persevere in solving
             them. Share and critique peer solutions.

©2015 Great Minds eureka-math.org
G5-M6-SE-BK3-1.3.1-02.2016

151

The following problems are puzzles for your enjoyment. They are intended to encourage working together and family problem-solving fun and are not a required element of this homework assignment.

2. Take 12 matchsticks arranged in a grid as shown below, and remove 2 matchsticks so 2 squares remain. How can you do this? Draw the new arrangement.

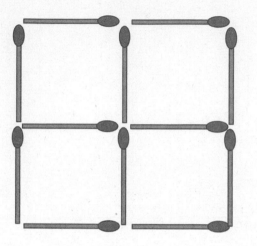

3. Moving only 3 matchsticks makes the fish turn around and swim the opposite way. Which matchsticks did you move? Draw the new shape.

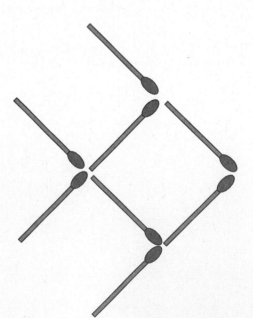

Lesson 23:   Make sense of complex, multi-step problems, and persevere in solving them. Share and critique peer solutions.

©2015 Great Minds eureka-math.org
G5-M6-SE-BK3-1.3.1-02.2016

Name _____   Date _____

1. Pat's Potato Farm grew 490 pounds of potatoes. Pat delivered $\frac{3}{7}$ of the potatoes to a vegetable stand.
    The owner of the vegetable stand delivered $\frac{2}{3}$ of the potatoes he bought to a local grocery store, which
    packaged half of the potatoes that were delivered into 5-pound bags. How many 5-pound bags did the
    grocery store package?

Lesson 24:   Make sense of complex, multi-step problems, and persevere in solving
               them. Share and critique peer solutions.

©2015 Great Minds eureka-math.org
G5-M6-SE-BK3-1.3.1-02.2016

153

The following problems are for your enjoyment. They are intended to encourage working together and family problem-solving fun. They are not a required element of this homework assignment.

2. Six matchsticks are arranged into an equilateral triangle. How can you arrange them into 4 equilateral triangles without breaking or overlapping any of them? Draw the new shape.

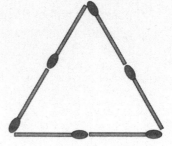

3. Kenny's dog, Charlie, is really smart! Last week, Charlie buried 7 bones in all. He buried them in 5 straight lines and put 3 bones in each line. How is this possible? Sketch how Charlie buried the bones.

Lesson 24: Make sense of complex, multi-step problems, and persevere in solving them. Share and critique peer solutions.

©2015 Great Minds eureka-math.org
G5-M6-SE-BK3-1.3.1-02.2016

Name _____   Date _____

1. Fred and Ethyl had 132 flowers altogether at first. After Fred sold $\frac{1}{4}$ of his flowers and Ethyl sold 48 of her flowers, they had the same number of flowers left. How many flowers did each of them have at first?

        Lesson 25:   Make sense of complex, multi-step problems and persevere in solving   155
                                them. Share and critique peer solutions.

©2015 Great Minds eureka-math.org
G5-M6-SE-BK3-1.3.1-02.2016

The following problems are puzzles for your enjoyment. They are intended to encourage working together and family problem-solving fun. They are not a required element of this homework assignment.

2. Without removing any, move 2 matchsticks to make 4 identical squares. Which matchsticks did you move? Draw the new shape.

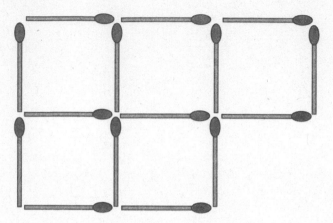

3. Move 3 matchsticks to form exactly (and only) 3 identical squares. Which matchsticks did you move? Draw the new shape.

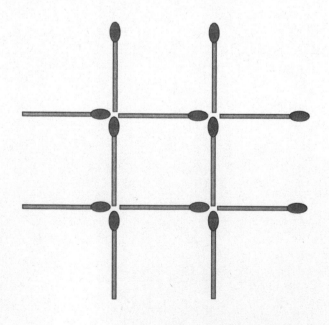

Lesson 25: Make sense of complex, multi-step problems and persevere in solving them. Share and critique peer solutions.

©2015 Great Minds eureka-math.org
G5-M6-SE-BK3-1.3.1-02.2016

Name _____ Date _____

1. For each written phrase, write a numerical expression, and then evaluate your expression.

   a. Three fifths of the sum of thirteen and six

      Numerical expression:

      Solution:

   b. Subtract four thirds from one seventh of sixty-three.

      Numerical expression:

      Solution:

   c. Six copies of the sum of nine fifths and three

      Numerical expression:

      Solution:

   d. Three fourths of the product of four fifths and fifteen

      Numerical expression:

      Solution:

2. Write at least 2 numerical expressions for each phrase below. Then, solve.

    a. Two thirds of eight

    b. One sixth of the product of four and nine

3. Use <, >, or = to make true number sentences without calculating. Explain your thinking.

    a. $217 \times (42 + \frac{48}{5})$ $\bigcirc$ $(217 \times 42) + \frac{48}{5}$

    b. $(687 \times \frac{3}{16}) \times \frac{7}{12}$ $\bigcirc$ $(687 \times \frac{3}{16}) \times \frac{3}{12}$

    c. $5 \times 3.76 + 5 \times 2.68$ $\bigcirc$ 5×6.99

Lesson 26:   Solidify writing and interpreting numerical expressions.

Name _____ Date _____

1. For each written phrase, write a numerical expression, and then evaluate your expression.

   a. Forty times the sum of forty-three and fifty-seven

      Numerical expression:

      Solution:

   b. Divide the difference between one thousand three hundred and nine hundred fifty by four.

      Numerical expression:

      Solution:

   c. Seven times the quotient of five and seven

      Numerical expression:

      Solution:

   d. One fourth the difference of four sixths and three twelfths

      Numerical expression:

      Solution:

2. Write at least 2 numerical expressions for each written phrase below. Then, solve.

   a. Three fifths of seven

   b. One sixth the product of four and eight

3. Use <, >, or = to make true number sentences without calculating. Explain your thinking.

   a.   4 tenths + 3 tens + 1 thousandth $\bigcirc$ 30.41

   b. $(5 \times \frac{1}{10}) + (7 \times \frac{1}{1000})$ 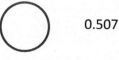   0.507

   c.   8 × 7.20 $\bigcirc$ 8 × 4.36 + 8 × 3.59

Lesson 26:   Solidify writing and interpreting numerical expressions.

six sevenths of nine	two thirds the sum of twenty-three and fifty-seven	forty-three less than three fifths of the product of ten and twenty	five sixths the difference of three hundred twenty-nine and two hundred eighty-one
three times as much as the sum of three fourths and two thirds	the difference between thirty thirties and twenty-eight thirties	twenty-seven more than half the sum of four and one eighth and six and two thirds	the sum of eighty-eight and fifty-six divided by twelve
the product of nine and eight divided by four	one sixth the product of twelve and four	six copies of the sum of six twelfths and three fourths	double three fourths of eighteen

expression cards

Lesson 26:   Solidify writing and interpreting numerical expressions.

©2015 Great Minds eureka-math.org
G5-M6-SE-BK3-1.3.1-02.2016

This page intentionally left blank

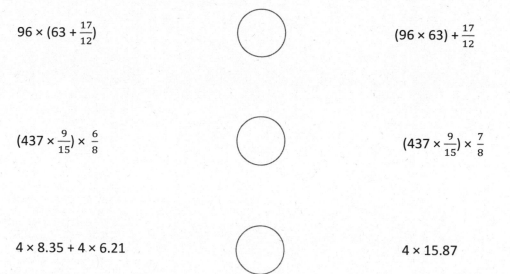

$96 \times (63 + \frac{17}{12})$ ◯ $(96 \times 63) + \frac{17}{12}$

$(437 \times \frac{9}{15}) \times \frac{6}{8}$ ◯ $(437 \times \frac{9}{15}) \times \frac{7}{8}$

$4 \times 8.35 + 4 \times 6.21$ ◯ 4×15.87

$\frac{6}{7} \times (3{,}065 + 4{,}562)$ ◯ $(3{,}065 + 4{,}562) + \frac{6}{7}$

$(8.96 \times 3) + (5.07 \times 8)$ ◯ $(8.96 + 3) \times (5.07 + 8)$

$(297 \times \frac{16}{15}) + \frac{8}{3}$ ◯ $(297 \times \frac{13}{15}) + \frac{8}{3}$

$\frac{12}{7} \times (\frac{5}{4} + \frac{5}{9})$ ◯ $\frac{12}{7} \times \frac{5}{4} + \frac{12}{7} \times \frac{5}{9}$

comparing expressions game board

EUREKA MATH

Lesson 26:   Solidify writing and interpreting numerical expressions.   163

©2015 Great Minds eureka-math.org
G5-M6-SE-BK3-1.3.1-02.2016

This page intentionally left blank

Name _____   Date _____

1. Use the RDW process to solve the word problems below.

    a. Julia completes her homework in an hour. She spends $\frac{7}{12}$ of the time doing her math homework and $\frac{1}{6}$ of the time practicing her spelling words. The rest of the time she spends reading. How many minutes does Julia spend reading?

    b. Fred has 36 marbles. Elise has $\frac{8}{9}$ as many marbles as Fred. Annika has $\frac{3}{4}$ as many marbles as Elise. How many marbles does Annika have?

2. Write and solve a word problem that might be solved using the expressions in the chart below.

Expression	Word Problem	Solution
$\dfrac{2}{3} \times 18$		
$(26 + 34) \times \dfrac{5}{6}$		
$7 - \left(\dfrac{5}{12} + \dfrac{1}{2}\right)$		

Lesson 27:   Solidify writing and interpreting numerical expressions.

Name _____   Date _____

1. Use the RDW process to solve the word problems below.

   a. There are 36 students in Mr. Meyer's class. Of those students, $\frac{5}{12}$ played tag at recess, $\frac{1}{3}$ played kickball, and the rest played basketball. How many students in Mr. Meyer's class played basketball?

   b. Julie brought 24 apples to school to share with her classmates. Of those apples, $\frac{2}{3}$ are red, and the rest are green. Julie's classmates ate $\frac{3}{4}$ of the red apples and $\frac{1}{2}$ of the green apples. How many apples are left?

2. Write and solve a word problem for each expression in the chart below.

Expression	Word Problem	Solution
$144 \times \dfrac{7}{12}$		
$9 - \left(\dfrac{4}{9} + \dfrac{1}{3}\right)$		
$\dfrac{3}{4} \times (36 + 12)$		

Name _____ Date _____

1. Answer the following questions about fluency.

   a. What does being fluent with a math skill mean to you?

   b. Why is fluency with certain math skills important?

   c. With which math skills do you think you should be fluent?

   d. With which math skills do you feel most fluent? Least fluent?

   e. How can you continue to improve your fluency?

2. Use the chart below to list skills from today's activities with which you are fluent.

Fluent Skills

3. Use the chart below to list skills we practiced today with which you are less fluent.

Skills to Practice More

Lesson 28: Solidify fluency with Grade 5 skills.

Name _____ Date _____

1. Use what you learned about your fluency skills today to answer the questions below.

   a. Which skills should you practice this summer to maintain and build your fluency? Why?

   b. Write a goal for yourself about a skill that you want to work on this summer.

   c. Explain the steps you can take to reach your goal.

   d. How will reaching this goal help you as a math student?

2. In the chart below, plan a new fluency activity that you can play at home this summer to help you build or maintain a skill that you listed in Problem 1(a). When planning your activity, be sure to think about the factors listed below:

- ▪ The materials that you'll need.
- ▪ Who can play with you (if more than 1 player is needed).
- ▪ The usefulness of the activity for building your skills.

Skill:	
Name of Activity:	
Materials Needed:	
Description:	

Lesson 28:   Solidify fluency with Grade 5 skills.

©2015 Great Minds eureka-math.org
G5-M6-SE-BK3-1.3.1-02.2016

Write Fractions as Mixed Numbers

Materials: (S) Personal white board

T: (Write $\frac{13}{2}$ = ____ ÷ ____ = ____.) Write the fraction as a division problem and mixed number.

S: (Write $\frac{13}{2}$ = 13 ÷ 2 = $6\frac{1}{2}$.)

More practice!

$\frac{11}{2}$, $\frac{17}{2}$, $\frac{44}{2}$, $\frac{31}{10}$, $\frac{23}{10}$, $\frac{47}{10}$, $\frac{89}{10}$, $\frac{8}{3}$, $\frac{13}{3}$, $\frac{26}{3}$, $\frac{9}{4}$, $\frac{13}{4}$, $\frac{15}{4}$, and $\frac{35}{4}$.

Fraction of a Set

Materials: (S) Personal white board

T: (Write $\frac{1}{2}$ × 10.) Draw a tape diagram to model the whole number.

S: (Draw a tape diagram, and label it 10.)

T: Draw a line to split the tape diagram in half.

S: (Draw a line.)

T: What is the value of each part of your tape diagram?

S: 5.

T: So, what is $\frac{1}{2}$ of 10?

S: 5.

More practice!

$8 \times \frac{1}{2}$, $8 \times \frac{1}{4}$, $6 \times \frac{1}{3}$, $30 \times \frac{1}{6}$, $42 \times \frac{1}{7}$, $42 \times \frac{1}{6}$, $48 \times \frac{1}{8}$, $54 \times \frac{1}{9}$, and $54 \times \frac{1}{6}$.

Convert to Hundredths

Materials: (S) Personal white board

T: (Write $\frac{3}{4} = \frac{}{100}$.) 4 times what factor equals 100?

S: 25.

T: Write the equivalent fraction.

S: (Write $\frac{3}{4} = \frac{75}{100}$.)

More practice!

$\frac{3}{4} = \frac{}{100}$, $\frac{1}{50} = \frac{}{100}$, $\frac{3}{50} = \frac{}{100}$, $\frac{1}{20} = \frac{}{100}$, $\frac{3}{20} = \frac{}{100}$, $\frac{1}{25} = \frac{}{100}$, and $\frac{2}{25} = \frac{}{100}$.

Multiply a Fraction and a Whole Number

Materials: (S) Personal white board

T: (Write $\frac{8}{4}$.) Write the corresponding division sentence.

S: (Write 8 ÷ 4 = 2.)

T: (Write $\frac{1}{4}$ × 8.) Write the complete multiplication sentence.

S: (Write $\frac{1}{4}$ × 8 = 2.)

More practice!

$\frac{18}{6}$, $\frac{15}{3}$, $\frac{18}{3}$, $\frac{27}{9}$, $\frac{54}{6}$, $\frac{51}{3}$, and $\frac{63}{7}$.

fluency activities

Lesson 28: Solidify fluency with Grade 5 skills.

©2015 Great Minds eureka-math.org
G5-M6-SE-BK3-1.3.1-02.2016

This page intentionally left blank

Multiply Mentally

Materials: (S) Personal white board

T: (Write 9 × 10.) On your personal white board, write the complete multiplication sentence.

S: (Write 9 × 10 = 90.)

T: (Write 9 × 9 = 90 – ____ below 9 × 10 = 90.) Write the number sentence, filling in the blank.

S: (Write 9 × 9 = 90 – 9.)

T: 9 × 9 is...?

S: 81.

More practice!

9 × 99, 15 × 9, and 29 × 99.

Find the Product

Materials: (S) Personal white board

T: (Write 4 × 3.) Complete the multiplication sentence giving the second factor in unit form.

S: (Write 4 × 3 ones = 12 ones.)

T: (Write 4 × 0.2.) Complete the multiplication sentence giving the second factor in unit form.

S: (Write 4 × 2 tenths = 8 tenths.)

T: (Write 4 × 3.2.) Complete the multiplication sentence giving the second factor in unit form.

S: (Write 4 × 3 ones 2 tenths = 12 ones 8 tenths.)

T: Write the complete multiplication sentence.

S: (Write 4 × 3.2 = 12.8.)

More practice!

4 × 3.21, 9 × 2, 9 × 0.1, 9 × 0.03, 9 × 2.13, 4.012 × 4, and 5 × 3.2375.

One Unit More

Materials: (S) Personal white board

T: (Write 5 tenths.) On your personal white board, write the decimal that's one-tenth more than 5 tenths.

S: (Write 0.6.)

More practice!

5 hundredths, 5 thousandths, 8 hundredths, and 2 thousandths. Specify the unit of increase.

T: (Write 0.052.) Write one more thousandth.

S: (Write 0.053.)

More practice!

1 tenth more than 35 hundredths,
1 thousandth more than 35 hundredths, and
1 hundredth more than 438 thousandths.

Add and Subtract Decimals

Materials: (S) Personal white board

T: (Write 7 ones + 258 thousandths + 1 hundredth = ____.) Write the addition sentence in decimal form.

S: (Write 7 + 0.258 + 0.01 = 7.268.)

More practice!

7 ones + 258 thousandths + 3 hundredths,
6 ones + 453 thousandths + 4 hundredths,
2 ones + 37 thousandths + 5 tenths, and
6 ones + 35 hundredths + 7 thousandths.

T: (Write 4 ones + 8 hundredths – 2 ones = ____ ones ____ hundredths.) Write the subtraction sentence in decimal form.

S: (Write 4.08 – 2 = 2.08.)

More practice!

9 tenths + 7 thousandths – 4 thousandths,
4 ones + 582 thousandths – 3 hundredths,
9 ones + 708 thousandths – 4 tenths, and
4 ones + 73 thousandths – 4 hundredths.

fluency activities

This page intentionally left blank

Decompose Decimals

Materials: (S) Personal white board

T: (Project 7.463.) Say the number.
S: 7 and 463 thousandths.
T: Represent this number in a two-part number bond with ones as one part and thousandths as the other part.
S: (Draw.)
T: Represent it again with tenths and thousandths.
S: (Draw.)
T: Represent it again with hundredths and thousandths.

More practice!

8.972 and 6.849.

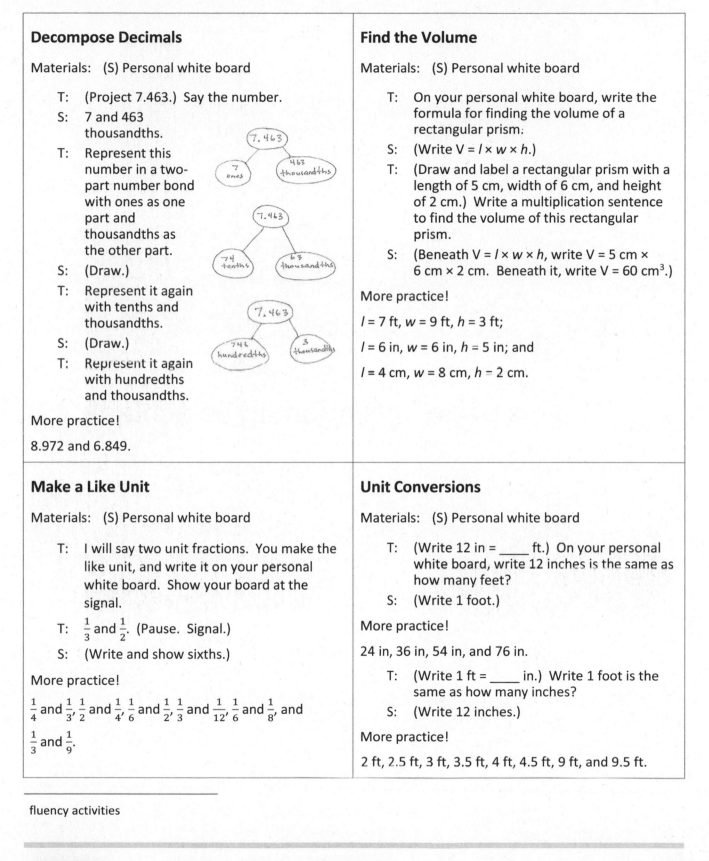

Find the Volume

Materials: (S) Personal white board

T: On your personal white board, write the formula for finding the volume of a rectangular prism.
S: (Write $V = l \times w \times h$.)
T: (Draw and label a rectangular prism with a length of 5 cm, width of 6 cm, and height of 2 cm.) Write a multiplication sentence to find the volume of this rectangular prism.
S: (Beneath $V = l \times w \times h$, write $V = 5$ cm $\times$ 6 cm $\times$ 2 cm. Beneath it, write $V = 60$ cm^3.)

More practice!

$l = 7$ ft, $w = 9$ ft, $h = 3$ ft;

$l = 6$ in, $w = 6$ in, $h = 5$ in; and

$l = 4$ cm, $w = 8$ cm, $h = 2$ cm.

Make a Like Unit

Materials: (S) Personal white board

T: I will say two unit fractions. You make the like unit, and write it on your personal white board. Show your board at the signal.
T: $\frac{1}{3}$ and $\frac{1}{2}$. (Pause. Signal.)
S: (Write and show sixths.)

More practice!

$\frac{1}{4}$ and $\frac{1}{3}$, $\frac{1}{2}$ and $\frac{1}{4}$, $\frac{1}{6}$ and $\frac{1}{2}$, $\frac{1}{3}$ and $\frac{1}{12}$, $\frac{1}{6}$ and $\frac{1}{8}$, and

$\frac{1}{3}$ and $\frac{1}{9}$.

Unit Conversions

Materials: (S) Personal white board

T: (Write 12 in = _____ ft.) On your personal white board, write 12 inches is the same as how many feet?
S: (Write 1 foot.)

More practice!

24 in, 36 in, 54 in, and 76 in.

T: (Write 1 ft = _____ in.) Write 1 foot is the same as how many inches?
S: (Write 12 inches.)

More practice!

2 ft, 2.5 ft, 3 ft, 3.5 ft, 4 ft, 4.5 ft, 9 ft, and 9.5 ft.

---

fluency activities

Lesson 28: Solidify fluency with Grade 5 skills.

177

©2015 Great Minds eureka-math.org
G5-M6-SE-BK3-1.3.1-02.2016

This page intentionally left blank

Compare Decimal Fractions

Materials: (S) Personal white board

T: (Write 13.78 ____ 13.86.) On your personal white board, compare the numbers using the greater than, less than, or equal sign.

S: (Write 13.78 < 13.86.)

More practice!

0.78 ____ $\frac{78}{100}$, 439.3 ____ 4.39, 5.08 ____ fifty-eight tenths, and thirty-five and 9 thousandths ____ 4 tens.

Round to the Nearest One

Materials: (S) Personal white board

T: (Write 3 ones 2 tenths.) Write 3 ones and 2 tenths as a decimal.

S: (Write 3.2.)

T: (Write 3.2 ≈ ____.) Round 3 and 2 tenths to the nearest whole number.

S: (Write 3.2 ≈ 3.)

More practice!

3.7, 13.7, 5.4, 25.4, 1.5, 21.5, 6.48, 3.62, and 36.52.

Multiplying Fractions

Materials: (S) Personal white board

T: (Write $\frac{1}{2} \times \frac{1}{3} =$ ____.) Write the complete multiplication sentence.

S: (Write $\frac{1}{2} \times \frac{1}{3} = \frac{1}{6}$.)

T: (Write $\frac{1}{2} \times \frac{3}{4} =$ ____.) Write the complete multiplication sentence.

S: (Write $\frac{1}{2} \times \frac{3}{4} = \frac{3}{8}$.)

T: (Write $\frac{2}{5} \times \frac{2}{3} =$ ____.) Write the complete multiplication sentence.

S: (Write $\frac{2}{5} \times \frac{2}{3} = \frac{4}{15}$.)

More practice!

$\frac{1}{2} \times \frac{1}{5}, \frac{1}{2} \times \frac{3}{5}, \frac{3}{4} \times \frac{3}{5}, \frac{4}{5} \times \frac{2}{3}$, and $\frac{3}{4} \times \frac{5}{6}$.

Divide Whole Numbers by Unit Fractions

Materials: (S) Personal white board

T: (Write $1 \div \frac{1}{2}$.) How many halves are in 1?

S: 2.

T: (Write $1 \div \frac{1}{2} = 2$. Beneath it, write $2 \div \frac{1}{2}$.) How many halves are in 2?

S: 4.

T: (Write $2 \div \frac{1}{2} = 4$. Beneath it, write $3 \div \frac{1}{2}$.) How many halves are in 3?

S: 6.

T: (Write $3 \div \frac{1}{2} = 6$. Beneath it, write $7 \div \frac{1}{2}$.) Write the complete division sentence.

S: (Write $7 \div \frac{1}{2} = 14$.)

More practice!

$1 \div \frac{1}{3}, 2 \div \frac{1}{5}, 9 \div \frac{1}{4}$, and $3 \div \frac{1}{8}$.

fluency activities

©2015 Great Minds eureka-math.org
G5-M6-SE-BK3-1.3.1-02.2016

This page intentionally left blank

Name _____ Date _____

1. Use your ruler, protractor, and set square to help you give as many names as possible for each figure below. Then, explain your reasoning for how you named each figure.

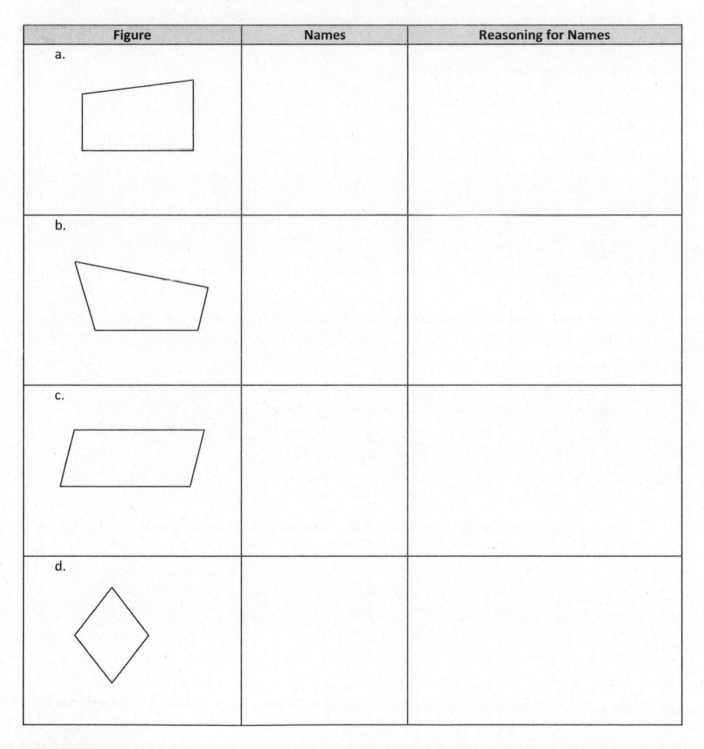

Figure	Names	Reasoning for Names
a.		
b.		
c.		
d.		

2. Mark draws a figure that has the following characteristics:

    ▪   Exactly 4 sides that are each 7 centimeters long.

    ▪   Two sets of parallel lines.

    ▪   Exactly 4 angles that measure 35 degrees, 145 degrees, 35 degrees, and 145 degrees.

    a. Draw and label Mark's figure below.

    b. Give as many names of quadrilaterals as possible for Mark's figure. Explain your reasoning for the names of Mark's figure.

    c. List the names of Mark's figure in Problem 2(b) in order from least specific to most specific. Explain your thinking.

Lesson 29:   Solidify the vocabulary of geometry.

©2015 Great Minds eureka-math.org
G5-M6-SE-BK3-1.3.1-02.2016

A quadrilateral with two pairs of equal sides that are also adjacent.	An angle that turns through $\frac{1}{360}$ of a circle.	A quadrilateral with at least one pair of parallel lines.	A closed figure made up of line segments.
Measurement of space or capacity.	A quadrilateral with opposite sides that are parallel.	An angle measuring 90 degrees.	The union of two different rays sharing a common vertex.
The number of square units that cover a two-dimensional shape.	Two lines in a plane that do not intersect.	The number of adjacent layers of the base that form a rectangular prism.	A three-dimensional figure with six square sides.
A quadrilateral with four 90-degree angles.	A polygon with 4 sides and 4 angles.	A parallelogram with all equal sides.	Cubes of the same size used for measuring.
Two intersecting lines that form 90-degree angles.	A three-dimensional figure with six rectangular sides.	A three-dimensional figure.	Any flat surface of a 3-D figure.
A line that cuts a line segment into two equal parts at 90 degrees.	Squares of the same size, used for measuring.	A rectangular prism with only 90-degree angles.	One face of a 3-D solid, often thought of as the surface upon which the solid rests.

geometry definitions

This page intentionally left blank

Base	Volume of a Solid	Cubic Units	Kite
Height	One-Degree Angle	Face	Trapezoid
Right Rectangular Prism	Perpendicular Bisector	Cube	Area
Perpendicular Lines	Rhombus	Parallel Lines	Angle
Polygon	Rectangular Prism	Parallelogram	Rectangle
Right Angle	Quadrilateral	Solid Figure	Square Units

geometry terms

©2015 Great Minds eureka-math.org
G5-M6-SE-BK3-1.3.1-02.2016

This page intentionally left blank

Name _____   Date _____

Teach someone at home how to play one of the games you played today with your pictorial vocabulary cards. Then, answer the questions below.

1. What games did you play?

2. Who played the games with you?

3. What was it like to teach someone at home how to play?

4. Did you have to teach the person who played with you any of the math concepts before you could play? Which ones? What was that like?

5. When you play these games at home again, what changes will you make? Why?

This page intentionally left blank

Attribute Buzz:

Number of players: 2

Description: Players place geometry terms cards facedown in a pile and, as they select cards, name the attributes of each figure within 1 minute.

- Player A flips the first card and says as many attributes as possible within 30 seconds.

- Player B says, "Buzz," when or if Player A states an incorrect attribute or time is up.

- Player B explains why the attribute is incorrect (if applicable) and can then start listing attributes about the figure for 30 seconds.

- Players score a point for each correct attribute.

- Play continues until students have exhausted the figure's attributes. A new card is selected, and play continues. The player with the most points at the end of the game wins.

Concentration:

Number of players: 2–6

Description: Players persevere to match term cards with their definition and description cards.

- Create two identical arrays side by side: one of term cards and one of definition and description cards.

- Players take turns flipping over pairs of cards to find a match. A match is a vocabulary term and its definition or description card. Cards keep their precise location in the array if not matched. Remaining cards are not reconfigured into a new array.

- After all cards are matched, the player with the most pairs is the winner.

Three Questions to Guess My Term!

Number of players: 2–4

Description: A player selects and secretly views a term card. Other players take turns asking yes or no questions about the term.

- Players can keep track of what they know about the term on paper.

- Only yes or no questions are allowed. ("What kind of angles do you have?" is not allowed.)

- A final guess must be made after 3 questions but may be made sooner. Once a player says, "This is my guess," no more questions may be asked by that player.

- If the term is guessed correctly after 1 or 2 questions, 2 points are earned. If all 3 questions are used, only 1 point is earned.

- If no player guesses correctly, the card holder receives the point.

- The game continues as the player to the card holder's left selects a new card and questioning begins again.

- The game ends when a player reaches a predetermined score.

Bingo:

Number of players: at least 4–whole class

Description: Players match definitions to terms to be the first to fill a row, column, or diagonal.

- Players write a geometry term in each box of the math bingo card. Each term should be used only once. The box that says *Math Bingo!* is a free space.

- Players place the filled-in math bingo template in their personal white boards.

- One person is the caller and reads the definition from a geometry definition card.

- Players cross off or cover the term that matches the definition.

- "Bingo!" is called when 5 vocabulary terms in a row are crossed off diagonally, vertically, or horizontally. The free space counts as 1 box toward the needed 5 vocabulary terms.

- The first player to have 5 in a row reads each crossed-off word, states the definition, and gives a description or an example of each word. If all words are reasonably explained as determined by the caller, the player is declared the winner.

game directions

This page intentionally left blank

		Math BINGO		

		Math BINGO		

bingo card

This page intentionally left blank

Name _____ Date _____

©2015 Great Minds eureka-math.org
G5-M6-SE-BK3-1.3.1-02.2016

This page intentionally left blank

Name _____ Date _____

1. List the Fibonacci numbers up to 21, and create, on the graph below, a spiral of squares corresponding to each of the numbers you write.

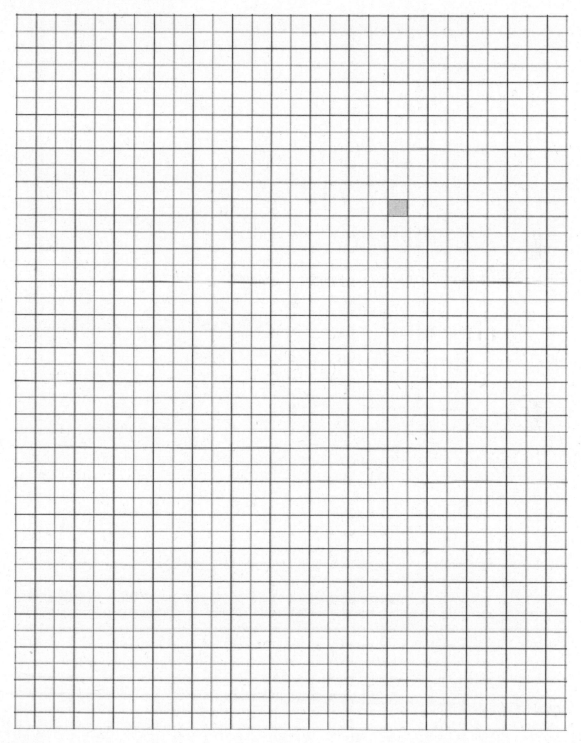

2. In the space below, write a rule that generates the Fibonacci sequence.

3. Write at least the first 15 numbers of the Fibonacci sequence.

EUREKA
MATH

Name _____ Date _____

1. Ashley decides to save money, but she wants to build it up over a year. She starts with $1.00 and adds 1 more dollar each week. Complete the table to show how much she will have saved after a year.

Week	Add	Total		Week	Add	Total
1	$1.00	$1.00		27		
2	$2.00	$3.00		28		
3	$3.00	$6.00		29		
4	$4.00	$10.00		30		
5				31		
6				32		
7				33		
8				34		
9				35		
10				36		
11				37		
12				38		
13				39		
14				40		
15				41		
16				42		
17				43		
18				44		
19				45		
20				46		
21				47		
22				48		
23				49		
24				50		
25				51		
26				52		

©2015 Great Minds eureka-math.org
G5-M6-SE-BK3-1.3.1-02.2016

2. Carly wants to save money, too, but she has to start with the smaller denomination of quarters. Complete the second chart to show how much she will have saved by the end of the year if she adds a quarter more each week. Try it yourself, if you can and want to!

Week	Add	Total		Week	Add	Total
1	$0.25	$0.25		27		
2	$0.50	$0.75		28		
3	$0.75	$1.50		29		
4	$1.00	$2.50		30		
5				31		
6				32		
7				33		
8				34		
9				35		
10				36		
11				37		
12				38		
13				39		
14				40		
15				41		
16				42		
17				43		
18				44		
19				45		
20				46		
21				47		
22				48		
23				49		
24				50		
25				51		
26				52		

Lesson 32:   Explore patterns in saving money.

3. David decides he wants to save even more money than Ashley did. He does so by adding the next Fibonacci number instead of adding $1.00 each week. Use your calculator to fill in the chart and find out how much money he will have saved by the end of the year. Is this realistic for most people? Explain your answer.

Week	Add	Total	Week	Add	Total
1	$1	$1	27		
2	$1	$2	28		
3	$2	$4	29		
4	$3	$7	30		
5	$5	$12	31		
6	$8	$20	32		
7			33		
8			34		
9			35		
10			36		
11			37		
12			38		
13			39		
14			40		
15			41		
16			42		
17			43		
18			44		
19			45		
20			46		
21			47		
22			48		
23			49		
24			50		
25			51		
26			52		

©2015 Great Minds eureka-math.org
G5-M6-SE-BK3-1.3.1-02.2016

This page intentionally left blank

Name _____   Date _____

1. Jonas played with the Fibonacci sequence he learned in class. Complete the table he started.

1	2	3	4	5	6	7	8	9	10
1	1	2	3	5	8				

11	12	13	14	15	16	17	18	19	20

2. As he looked at the numbers, Jonas realized he could play with them. He took two consecutive numbers in the pattern and multiplied them by themselves and then added them together. He found they made another number in the pattern. For example, $(3 \times 3) + (2 \times 2) = 13$, another number in the pattern. Jonas said this was true for any two consecutive Fibonacci numbers. Was Jonas correct? Show your reasoning by giving at least two examples of why he was or was not correct.

3. Fibonacci numbers can be found in many places in nature, for example, the number of petals in a daisy, the number of spirals in a pine cone or a pineapple, and even the way branches grow on a tree. Find an example of something natural where you can see a Fibonacci number in action, and sketch it here.

This page intentionally left blank

Name _____ Date _____

Record the dimensions of your boxes and lid below. Explain your reasoning for the dimensions you chose for Box 2 and the lid.

BOX 1 (Can hold Box 2 inside.)

The dimensions of Box 1 are _____ × _____ × _____ .

Its volume is _____ .

BOX 2 (Fits inside of Box 1.)

The dimensions of Box 2 are _____ × _____ × _____ .

Reasoning:

LID (Fits snugly over Box 1 to protect the contents.)

The dimensions of the lid are _____ × _____ × _____ .

Reasoning:

EUREKA
MATH™

Lesson 33: Design and construct boxes to house materials for summer use.

203

©2015 Great Minds eureka-math.org
G5-M6-SE-BK3-1.3.1-02.2016

1. What steps did you take to determine the dimensions of the lid?

2. Find the volume of Box 2. Then, find the difference in the volumes of Boxes 1 and 2.

3. Imagine Box 3 is created such that each dimension is 1 cm less than that of Box 2. What would the volume of Box 3 be?

©2015 Great Minds eureka-math.org
G5-M6-SE-BK3-1.3.1-02.2016

Name _____   Date _____

1. Find various rectangular boxes at your home. Use a ruler to measure the dimensions of each box to the nearest centimeter. Then, calculate the volume of each box. The first one is partially done for you.

Item	Length	Width	Height	Volume
Juice Box	11 cm	2 cm	5 cm	

2. The dimensions of a small juice box are 11 cm by 4 cm by 7 cm. The super-size juice box has the same height of 11 cm but double the volume. Give two sets of the possible dimensions of the super-size juice box and the volume.

Lesson 33:   Design and construct boxes to house materials for summer use.

©2015 Great Minds eureka-math.org
G5-M6-SE-BK3-1.3.1-02.2016

This page intentionally left blank

Name _____ Date _____

I reviewed _____'s work.

Use the chart below to evaluate your friend's two boxes and lid. Measure and record the dimensions, and calculate the box volumes. Then, assess suitability, and suggest improvements in the adjacent columns.

Dimensions and Volume	Is the Box or Lid Suitable? Explain.	Suggestions for Improvement
BOX 1 dimensions: Total volume:		
BOX 2 dimensions: Total volume:		
LID dimensions:		

Lesson 34:   Design and construct boxes to house materials for summer use.

207

©2015 Great Minds eureka-math.org
G5-M6-SE-BK3-1.3.1-02.2016

This page intentionally left blank